Risk Perception of Artificial Intelligence

Risk Perception of Artificial Intelligence

Hugo Neri

LEXINGTON BOOKS
Lanham • Boulder • New York • London

Published by Lexington Books
An imprint of The Rowman & Littlefield Publishing Group, Inc.
4501 Forbes Boulevard, Suite 200, Lanham, Maryland 20706
www.rowman.com

6 Tinworth Street, London SE11 5AL, United Kingdom

British Library Cataloguing in Publication Information Available

Library of Congress Cataloging-in-Publication Data Available

Library of Congress Control Number: 2020948467

ISBN 978-1-7936-0205-3 (cloth)
ISBN 978-1-7936-0206-0 (electronic)

Contents

Preface

This research began during a short internship at the Knowledge Lab, University of Chicago, headed by Professor James Evans, in the Summer of 2017. At that moment, I was willing to model the common-sense representations people could have about artificial intelligence based on Twitter data because of its abundance of material available chronologically. I wanted to observe especially the change in the popular representations of artificial intelligence from the moment Twitter started to run to the moment of that research. In 2017, the term "artificial intelligence" was rising in popularity, along with some risk associations, mainly motivated by some public positionings from prominent scientists and businessmen, such as Stephen Hawking, Elon Musk, and Stuart Russell, to name a few. A new outlook of artificial intelligence (i.e., non-cliché) was also depicted in the movies, Spike Jonze's *Her* (2013), Alex Garland's *Ex Machina* (2014), and in an updated cyberpunk idea we find in Wally Pfister's *Transcendence* (2014). Before the 2010s, I had the impression that the term was disregarded as something not serious enough among the scholar community, something that belonged only to science fiction reveries. It also had a strong association with electronic games, especially the level of difficulty one could select. However, it has changed, and artificial intelligence grew in importance as one of the most (if not the most) relevant piece of technology under development.

As mentioned above, some key figures drove the new representation and the possible outcomes it could bring. What took my attention was that some experts, scientists, and intellectuals were coming to the public to talk about this technology and what could happen to us if we succeeded in having a full-fledged artificial intelligence. This last part is filled with undesired consequences that we would like to avoid if we could or to be prepared if there is nothing in our hands. It also came to my attention that there was a missing piece in the framework of risk when I wanted to consider the public perception of the risk of artificial intelligence. In the social amplification of risk framework (SARF) developed by Paul Slovic, Roger Kasperson, and others, the representations of risk were supported by real events (mishaps) that happened in the past. It would be the sparking fire for their amplification and economic, social, and political impacts. The missing piece was that there were no real events that could support the risk representations of artificial intelligence. However, they have been amplified. Some institutional effects could already be noticed, such as the creation of some institutes dedicated to study and propose safety measures for the development of artificial intelligence. Two of them are at the University of Cambridge, the Center for the Study of Existential Risk, and the Center for the Future of Intelligence—we should also mention the Future of Life Institute at the University of Oxford. In addition to that, the head of the Department of Sociology of the University of Cambridge at that moment was Patrick Baert, who has been developing a theory on public intellectuals. It seemed to be a place to be for developing this research. So, during 2018, I was a visiting scholar at the University of Cambridge, and Professor Baert was my host.

Besides the centers that were conducting serious research, there was also happening at the University of Cambridge some religious debates on the consequences of artificial intelligence for human existence. The key institution promoting a cycle of debates and public lectures was the Faraday Institute. By joining the events, I could realize how deeply artificial intelligence could affect a set of metaphysical and cultural conceptions of the role of humankind. As a Christian institute, the Christian system of belief was at the play. However, I had the luck to meet a scholar, Yaqub Chaudhury, from the Cambridge Muslim College, who shared a valuable assessment of the tensions that the full-fledged artificial intelligence with the Muslim system of beliefs. On this religious front, I interviewed two more scholars, Michael Burdett (University of Nottingham), a philosophy and theology professor at Oxford, and David Glass, a computer scientist professor at Ulster University with a deep interest in the science and religion interaction.

In the first semester of 2018, there was an Artificial Intelligence Committee promoted by House of Lords in which many players from the industry and academia positioned themselves publicly regarding the role of artificial intelligence for the future of the UK economically and socially speaking, as well as its potential threats. I interviewed three scholars who took part in the public session of the committee, Professor Sir David Spiegelhalter University of Cambridge), Professor Peter McOwan (QueenMary, University of London), who unfortunately died last year, and Professor David Edgerton. It is also worthy to note that Professor McOwan emphasized the cultural difference between Japan and "the West" in the representation of artificial intelligence, something we have considered in this book. Professor Edgerton also portrayed a historical approach and especially the role of futurists. Personally, for those who appreciate a social history of technology, his book *The Shock of the Old* is highly recommended. It was also during this period that the Cambridge Analytica's scandal with Facebook came to the public. Even though it was not directly related to artificial intelligence, the issue of data misuse raised some concerns on how machine learning models have been trained. It was also during this period that the Cambridge Analytica's scandal with Facebook came to the public. It certainly made the public more aware of the practical risks of artificial intelligence.

I had fifteen interviewees in total, some I talked more than once. All of them provided significant insights into the development of the research and this book. I also would like to mention professor Jeremy Wyatt of the University of Birmingham, who emphasized the need for developing artificial intelligence, especially robots, which would ease the burden of aging and care. Professor Jose Hernández-Orallo (Universitat Politecnica de Valencia) and Professor Aaron Sloman (University of Birmingham) raised questions on what intelligence is. Since the beginning of the discipline, Professor Sloman is a legendary figure in the field, mainly because of his philosophical interests with artificial intelligence. Marvin Minsky used to recommend to his students the work of Sloman, as "he is a philosopher who knows how to code." Professor Wolfgang Coy (Humboldt-Universität) shared much of professor Sloman's insights as an old member of the artificial intelligence community. It is also interesting to mention the interview with David Levy, who was a former chess champion, a businessman, also known for his book on a sexual relationship with robots. Even though this book is not about those interviews, they provided crucial insights into how I should address this topic of the risk perception of artificial intelligence.

Professor Baert held meetings with his group to discuss the members' ongoing research. When presenting my working paper, I received many valuable contributions. Among all the comments, I would like to acknowledge Eliran Bar-El for making me paying particular attention to the movie's representation. At that time, Blade Runner 2049 was released a year before. The contact with professor Baert and his group was a rich source of ideas on the intellectuals' role. Still about this group, professor Filipe Carreira da Silva (University of Cambridge) drew my attention to "the future," when discussing his interests in (political) utopia.

I also presented an early first version of this research at the 113th Annual Meeting of the American Sociological Association, in Philadelphia, in a Science, Knowledge, and Technology session entirely dedicated to artificial intelligence. A room full of sociologists on the last day of the congress suggested that Social Sciences and Humanities intensively discuss this topic.

During the congress, I got the invitation from Lexington to publish the research as a book, when I met Courtney Morales, the editor of this book, to whom I acknowledge the attention and patience with the development of the manuscript of this book.

During 2019, I started to work at the Center for Artificial Intelligence, a joint endeavor of the IBM and São Paulo Research Foundation, placed at the University of São Paulo. There, I have been researching on common sense reasoning and knowledge. Because of that, I could focus on the development of two important parts of the argument that were left, culture (including popular culture) and the future. Regarding the cultural aspect, I had a chance to have a good exchange of ideas with Dr. Veridiana Domingos Cordeiro, which became a collaborator in two chapters of this book. She explored and developed further the arguments about the Japanese societal and cultural relationships with intelligent machines. It ended up as an alternative development of the representation of artificial intelligence, which highlighted what was going on in the West front.

The final piece added to the work was the counterfactual thinking and its role in conceiving future scenarios. This idea entered unexpectedly in the book as I was developing research on common sense with Professor Fabio Cozman, director of the Center for Artificial Intelligence. It was motivated by our discussion on explainable artificial intelligence. Here, three different sources came together in the concept developed in this book, the work of Judea Pearl on causality, Daniel Kahneman on the psychology of counterfactual thinking, and David Lewis on truth in fiction.

This book is a string of reflections that compose a possible narrative on how our perception of artificial intelligence, this multilayered term, has been formed. Reflecting on such narrative amid cultural representation, technological innovations, and the role of experts is a multifaceted reading on the topic that will remain in the sights of the world for many years to come. I hope this book will help the readers to build a historical scenario about what has happened so far, so that they can continue to follow the development of artificial intelligence and how to understand it.

Introduction

Stories of the Future

Artificial intelligence's herald has been "the computer" HAL 9000 since its inception. The fact that HAL is the piece of technology that enables space travel in Stanley Kubrick's *2001: A Space Odyssey* captures the new possibilities opened as a consequence of having a "not-exactly human kind" of intelligence. But it also shows that among the new possibilities, some of them may include undesired consequences. And "risk" is narrowly contained in such unfavorable scenarios. In a broader sense, risk is related to the very adventure—or in this case, to the very odyssey. So, we have a cliffhanger idea related to risk. The representation of this idea of risk in a narrative is not exclusive of Kubrick's movie. Still, it may appear whenever we start to imagine the resulting scenarios as soon as we can conceive machines with intelligence or artificial intelligence.

In the long term, few events shape the representations of the risks of artificial intelligence. They can range from fictional stories to technological advancements. Still, the major events are surrounded or backed by a variety of minor accounts, declarations, and achievements. Among the illustrious minor unknown events, the following old story traces back to the beginning of the representations of machine intelligence, which still sounds contemporary for us. The narrative is a history of the future, written by an anonymous author for the first real science fiction venue, the pulp magazine *Amazing Stories,* owned by Hugo Gernsback's, in February of 1927. Although we will see in more detail what are these venues and their significance for the representation of the risks of artificial intelligence in chapter 2, let us begin with the story *The Though Machine* by *Ammianus Marcellinus*[1]—the author whose identity is unknown.

The story starts with a warning in a biblical form on the potential threat that the scientific knowledge may bring, "of all the trees in the garden of science man may eat, but of this tree he shall not eat, or he shall surely die."[2] Unsurprisingly, the forbidden fruit is artificial intelligence. Furthermore, the story is about how we got this fruit and then how we perish as humanity. The protagonist is Dr. Heinrich Schmidt, also known as Henry Smith in the United States, a German scientist and inventor who moved to New York, where he set his laboratory and became obsessed with an "almost impossible idea": building a thinking machine. It was necessary to have a protagonist who came from Germany to accomplish this goal since getting this holy grail required stereotypical German traits in the eyes of 1920s Americans: patience, hard-working, and lack of creativity. As a German, he was also a Nietzsche's reader and influenced by the *Übermensch*[3] idea. Nevertheless, he had to undertake his enterprise in America because it needed to be done as a business.

Throughout the story, Smith explores his approaches to the hard questions of replicating human's thinking in a machine. His lack of creativity became evident when he deals with the problem. For instance, "does it [the machine] perform the same operations as the human brain upon its problems?—no matter, since it gets the same results."[4] Also, "the human mind is a machine. There can be no doubt of that. . . . By dividing all possible thought into its simplest elements it should be possible to produce an external analogy to it in a machine."[5] Some correct guesses about the future consequently appeared in this action of imagining scenarios. The author correctly guesses that it would be possible to express ideas electrically—today, we would say electronically or digitally. Such predictions are always embedded in more complex scenarios that carry other beliefs, such as the triumph of philosophical functionalism. As a consequence, reproducing every single element of human intelligence is a task that requires both money and patience. Smith took all the money from the royalties of a previous invention, a "typewriter eraser," and know how to wait twenty years for completing the first model. In the beginning, there was no apparent use for the machine other than pure entertainment. Its first application was an odd attraction in the circus. However, it kept the money flowing in and allowed new developments to take place. Smith kept no secret regarding his invention. As a consequence, many have imitated his first model, and smaller machines have pop across the globe. Thus, the imitation honed and led to the perfection of the machines. The first model had no mind as Smith described it, even though it could perform several tasks at a human level. Regardless of the initial limitations, the machines were improved continuously and, at some point, they could improve themselves. It follows from this intelligence explosion that they could address any problem humans might have, not only physical but also intellectual.

Besides, for the next one hundred years, "machines did what was too strenuous for human muscles, and machines did what was too difficult for human brains. [. . .] To dream a thing was to accomplish it [. . .] they threw up their own questions like minds, and like minds answered them."[6] They were making constant scientific discoveries and building improved versions of themselves. Humanity had nothing to do but to sit under the palm trees and let progress unfold unremittingly. There was no reason or motivation to any physical or intellectual undertaking, as the author remarked, "a man who ventured to think was as ridiculous as a man who offered to dig with a shovel, while an excavating machine was standing by it."[7] By 2050, everything would be at an *optimum*. For humans, a life full of leisure is all that was left. This is the point when artificial intelligence brought unintended consequences. Humans became too dependent on the machines; and they became lazy, weak, and ignorant. At this moment, every piece of knowledge existed in the "electronic language," and no human needed to know it. Humanity was degenerating in such an earthly Eden. As soon as an unforeseen event destroyed the machines, humans fell once again into a primitive state.

This story conceives a chain of scenarios until the invention of intelligent machines, and then it imagines its consequences. From a Nietzschean narrative's standpoint, a society of leisure would inevitably lead to social degeneration. This conclusion, however, could vary across other imagined future worlds. It could be seen optimistically as a society without work where everyone could have access to goods. Different kinds of pessimistic scenarios could also be derived from this first chain of events: that the development of machines would not be left by themselves. Still, it would be under the control of a minority of machines' owners. From here, we would have a dystopian cyberpunk-like scenario in which inequality is a constitutive trait of such imagined worlds. The direction of these chains of scenarios depends on cultural influences.

Nonetheless, there is an invariant condition for having machines with intelligence, at least as intelligent as human beings: that they could quickly overcome humans. Here, "overcoming" does not necessarily imply that machines will take over humanity, but that human efforts to compete against a machine of this kind are so absurd as the legend of the folk hero John Henry in the late nineteenth century. John tried to save railroad workers' jobs threatened by drilling machines. As the legend goes, he competed and won against a steam-powered machine to decide who would drive in more spikes. Just after his victory, he had collapsed and died. As we can see, both stories, Marcellinus's *The Thought Machine* and the steel-driving man by John Henry, are cautionary tales about human limitations compared to machines. They tell us that our superseding is around the corner. Whatever the tone of the narrative about artificial intelligence is, from abstract systems to embodiment in robots, the replacement idea is inexorable.

EXPERTS ALSO CONCEIVE FUTURE RISKS

If we fast-forward about a hundred years from our anonymous author's story published in 1927, we may stumble upon a nonfiction public intervention regarding artificial intelligence made by a remarkable expert. The prestigious expert Stuart Russell conceived in his latest book (*Human Compatible: Artificial Intelligence and the Problem of Control*) that artificial intelligence could be humans' last invention, and further imagined what may happen if we succeed in this endeavor.[8] This idea of artificial intelligence as a "final invention" became popular in the last decade. We could also find many book titles expressing this very idea.[9] As we will see in chapter 5, there appeared a set of experts who advocated for the idea of imminent catastrophe linked to artificial intelligence. These advocators are renowned people, including the English theoretical physicist and cosmologist, Stephen Hawking. He said in 2014 in an interview for BBC that "the development of full artificial intelligence could spell the end of the human race."[10]

Moreover, in the published version of the same interview, HAL was selected to illustrate the future risk of artificial intelligence. In a picture of *2001's* scene, when HAL spied on the astronauts locked in a pod, the caption was saying that "[HAL] encapsulated many people's fears of how AI could pose a threat to human life." Contrary to these views, Russell disregarded such catastrophic positions once they are outside the field of artificial intelligence. However, Russell also makes his predictions from a nonfiction context. For example, in a talk for a lay audience regarding artificial intelligence in 2013, he nominated the five candidates for the biggest event of the future of humanity: (a) human extinction due to a variety of catastrophic events such as asteroid impacts, pandemics, and so on, (b) living forever, that is, the medical solution for aging; (c) the invention of faster than light travel and subsequently the conquer of the universe; (d) superior alien civilization visits us; finally (e) we invent superintelligent artificial intelligence. Among all the candidates, he believes that the fifth one is the most likely to happen. As he says, "the arrival of a super-intelligent AI is in many ways analogous to the arrival of a superior alien civilization but much more likely to occur."[11] It could be the last invention of human civilization, not because it will try to take over humans as soon as it becomes conscious, but mainly because of the misalignment of values between humans and machines. The problem of value alignment is the idea of having an intelligent being as a mere goal accomplisher. The problem lies in the machines' lack of what we call "common sense," so it could cause a set of unintended consequences simply because we cannot specify every scenario to avoid its absolute rational course of action. Besides the latent risk, we may still have many other explicit misuses of this technology, such as enhanced surveillance and autonomous weapons—as we will explore in chapter 5.

From the anonymous fiction writer to Russell's latest nonfiction book, we see this exercise of conceiving future scenarios. It may not only occur as the form of fiction or science fiction but as expressed beliefs from a variety of sources. The English historian David Edgerton claimed in his book *The Shock of the Old* that technological futurism is largely unchanging over time; it also displays a startling, unselfconscious lack of originality:

> We need to be aware that this futurology of the past has affected our history. From it we get our focus on invention and innovation, and on the technologies which we take to be the most important. From this literature, the work of low- and middle-ranking intellectuals and propagandists, ranging from, say, the books of H. G. Wells to the press releases of NASA's PR officials, we get a whole series of clichéd claims about technology and history. We should take them, not as well-grounded contributions to our understanding, for they rarely are that, but as the basis of questions. What have been the most significant technologies of the twentieth century? Has the world become a global village? Has culture lagged behind technology? Has technology had revolutionary or conservative social and political effects?[12]

This reflection serves to more comprehensive thinking about the actual development of artificial intelligence in the last decade, especially the beliefs, hopes, and fears it had infused and has been infusing in both experts and laypeople. As we present throughout the next chapters, the imagination of a machine that displays a full-fledged artificial intelligence and its associated risks started at the beginning of the twentieth century. Although some representations of the risk of artificial intelligence are new, many of them were brewed before. For this reason, we keep Edgerton's skepticism about technological forecasting to have a critical outlook on how this variety of risk came to be represented. In other words, how experts, writers, and futurologists in general thought about what can conceivably outgrowth from the invention of intelligent machines. Also, what are the dangerous fruits from this process?

Furthermore, we want to highlight that the comparison between experts' public positioning in nonfiction contexts and science fiction narratives does not intend to be derogatory. Such a comparison aims to show that conditional and counterfactual thinking is pervasive regarding technologies in general and under-construction technologies in particular. It eventuates our beliefs, hopes, and fears of future scenarios by extrapolating some known pieces of evidence in the present and reasoning about its consequences. Before we proceed, we want to contextualize the coinage of the term artificial intelligence.

ARTIFICIAL INTELLIGENCE AS A MAGNET TERM

John McCarthy, one of the founding fathers of the discipline, coined the term "Artificial Intelligence" when he and Claude Shannon, the author of the mathematical theory of information, were trying to get money for a summer study in 1956. The "Summer Research Project on Artificial Intelligence" was a proposal submitted to the Rockefeller Foundation in August 1955 and after turned to be "The Dartmouth Conference," where few but prominent intellectuals joined. How we recognize today, all the founding fathers of the discipline had been there: McCarthy, Marvin Minsky, Allen Newell, and Herbert Simon. The historical books of the discipline commonly recognize the conference as its foundational event. Although it may be the case, one should not romanticize its role. McCarthy's recollections on the occasion, for instance, depicts a situation that people (at least in academia) are pretty much aware of,

> Anybody who was there was pretty stubborn about pursuing the ideas that he had before he came, nor was there, as far as I could see, any real exchange of ideas. People came for different periods. The idea was that everyone would agree to come for six weeks, and the people came for periods ranging from two days to the whole six weeks, so not everybody was there at once. It was a great disappointment to me because it really meant that we couldn't have regular meetings.[13]

The experts who took part in this event were not entirely convinced about the name "artificial intelligence." Newell and Simon preferred the neutral and non-catching term "complex information processing," Shannon thought the name was too "flashy." Although Minsky employed the term after that, he had adopted the adjective "resourceful" rather than intelligent or smart to designate a full-fledge intelligent machine since his book *The Emotion Machine*.[14] It is also incorrect to think that regardless of the term, what the foundational event brought was the synthesis of many different disciplines and schools of thought in a single new enterprise. For example, Minsky recognizes it in an interview, "I think we had the feeling that these ideas were beginning to become popular, and maybe that's a historical event. It wasn't true. It took another ten years before people could tolerate the idea of AI without thinking that it was funny and impossible."[15]

Moreover, during the next two decades of research with outstanding results, many of the researchers involved in it did not recognize their efforts under the term "artificial intelligence." For instance, Professor Roger Needham, a researcher from the Computer Science Department of the University of Cambridge stated that "Artificial Intelligence is a rather pernicious label to attach to a very mixed bunch of activities, and one could argue that the sooner we forget it, the better."[16] Arthur Samuel, a pioneer of computer gaming at Stanford University, who coined the sibling catching term "machine learning," said the "the word 'artificial' makes you think there's something kind of phony about this [. . .] or else it sounds like it's all artificial and there's nothing real about this work at all."[17] Despite that, precisely because the term was "too flashy," as Shannon stated, or as Newell pointed out, he adopted the term decades after, "So cherish the name artificial intelligence. It is a good name. Like all names of scientific fields, it will grow to become what its field comes to mean."[18] It served as a great slogan that could become not only a conceptual umbrella but also a potentially trendy term.

The term "artificial intelligence" was born as a quirky project that was successful enough to be popularized decades after, even though it was not consensual in the computer science community neither the origins of the term were clear for those who had adopted it. Not only the name was and still is expressive enough, but also its goal, as McCarthy stated, was flexible enough to bring together two different groups of researchers according to their approach and expectations to the future of the discipline. It is a magnet term. McCarthy put it in the latest updated version of his paper *What Is Artificial Intelligence*, artificial intelligence is "the science and engineering of making intelligent machines, especially smart computer programs. It is related to the similar task of using computers to understand human intelligence, but AI does not have to confine itself to biologically observable methods."[19] The choice of the word "intelligence" is what ensures the degree of flexibility of the term.

Throughout the history of the discipline, the artificial intelligence researchers or merely the experts showed two kinds of public positions; they were either pragmatists or futurists. For pragmatists, we mean these experts whenever they talk about the development of the discipline or technology. They do so technically; they talk about the normal research. For futurists, we mean those experts who came to the public and not only make predictions regarding the technological development of artificial intelligence but also they draw the consequential future scenarios derived this invention. In some sense, both are optimists regarding the fulfillment of the goal of the discipline: achieving human-level intelligent machines—it seems to be a legit goal to have from a scientific and engineering perspectives. Thus, every expert may have their private imagination about the future if the discipline succeeds in creating a full-fledged artificial intelligence. They may even share it with his or her inner circle of relationships. Notwithstanding such possible general behavior, only very few of them come to the public with a new representation of the scenarios of artificial intelligence and its perils.

We may also raise a question of why some experts opt to become public personalities or public intellectuals.[20] Still, it would not necessarily shed light on the representations of the risk of artificial intelligence.

The futurists' expectations are high. Sometimes they sound a little bit prophetical as we will see throughout this book. However, during the peaks of the discipline, the optimism was ubiquitous. For example, in the late

1960s and early 1970s, we could find claims such as Simons's "machines will be capable, within twenty years, of doing any work a man can do,"[21] or Minsky's, "within a generation . . . the problem of creating 'artificial intelligence' will substantially be solved,"[22] and "from three to eight years we will have a machine with the general intelligence of an average human being."[23]

During the 1970s, the expectations were very high, with some reason. From 1956 to 1973, artificial intelligence became a popular subject due to its collection of various achievements. As Stuart Russell and Peter Norvig argued in their handbook of AI, we see in this period the "golden age" of AI, when "it was astonishing whenever a computer did anything remotely clever."[24] However, it collected many disappointments, as well. Precisely because of the mixture of disappointments, popularity, and attempts to define the discipline, its enterprise, and leading figures, an event now known as the "Lighthill's Report" brought to the front everything said so far. We will see this episode in more detail in chapter 3.

Finally, over the past century, there have been a variety of terms that designate artificial intelligence or this idea that a machine can show what we would call an "intelligent human behavior," whatever it means. In this sense, robots will stand for a class of similar words such as androids, droids, automata, cyborgs, and so on. The underlying idea is that the robot may have some degrees of freedom of action instead of being controlled remotely or previously. In other words, it can interact with its environment and act correspondently.

CRAFTING A STORY FOR AI: ONE AMONG MANY

The bulk of the risks of artificial intelligence do not come from breaking news' accidents with the technology in the past, or its uses in war, but it comes from the future. The future drives our risks of artificial intelligence, more precisely, the future scenarios in which something undesirable may happen. Futurists play a crucial part in this job of conceiving future scenarios and consequences, either in fiction or in nonfiction environments. Writers take the first on, whereas experts engage with the latter.

To analyze and discuss that, we set our approach to risk and counterfactual thinking in chapter 1. The risk approach lies in a psychosocial framework called the "social amplification of risk." This framework derives from a long sequence of psychometric works on risk perception. It was born with the ideas of bounded rationality and heuristics from Herbert Simon as well as Kahneman and Tversky, respectively.

Also, there is a critical approach to traditional risk analysis, which is a frequentist approach. Even though the "social amplification of risk" approach does not explore in-depth culture factors influencing the perception and representation of risk, it is open for that. Nevertheless, it has some limitations to our analysis. These limitations regard the under-emphasized role of the experts and the naive realist approach to how the representation of risk occurs; also, there is no consideration of role of the internet and social networks services in the process of propagation of the representations of risk. Moreover, it has a severe limitation regarding a historical perspective. Thereby, we try to amend these limitations. Since the role of experts and fiction is key to understand the fears of artificial intelligence, we make a careful characterization of these elements in chapter 1. In the sequence, chapters 2, 3, 4, and 5 draw a historical perspective analyzing the tripod "technological achievements—experts representations of the future risk of artificial intelligence—the future risks drawn by science fiction." Chapter 2 depicts the images of those future risks before the establishment of the scientific discipline in the 1950s. During this period, we see the rise of the science fiction market with pulp magazines and movies such as Fritz Lang's *Metropolis*. In chapter 3, we analyze the exploration era of artificial intelligence from the establishment of the discipline to the "Lighthill's report," a commissioned report by the UK. At the peak of this period, the avatar for artificial intelligence was born, HAL 9000, in the Stanley Kubrick and Arthur C. Clarke's *2001: A Space Odyssey*. In chapter 4, we explore the 1980s and the 1990s, the new Japanese threat, the rise of both "cyberpunk culture" and the futurist experts, such as Hans Moravec and mainly Ray Kurzweil. Chapter 5 depicts the contemporary period of artificial intelligence when significant achievements appeared, such as IBM Deep Blue victory on the chess-master Garry Kasparov, IBM Watson's victory in the trivia game *Jeopardy!*, and Google DeepMind AlphaGo's triumph over the Go grandmaster Lee Sedol. The 2010s is the period when interaction via internet and content sharing matured. This period is also a watershed because experts could address their views about the future directly, without appealing to fictional contexts. Among different venues, such as books, news media, general conferences, it was on the TED Talks that they were publicly legitimated. Scholars became superstars. The new developments of neural networks backed this movement.

Finally, chapter 6 rethink the risk perception of artificial intelligence from another standpoint. Considering the relation between Japanese society and the representation of artificial intelligence in robots bodies, we discuss how the risk perception may be a Western cultural orientation. Also, by comparing drawings of robots, we puzzle over the (human and non-human) attributes people conceive when thinking on artificial intelligence.

NOTES

1. Marcellinus (329 CE–394 CE) was a real Roman historian who wrote the *Res Gestae*, a history of the accession of the emperor Nerva in 96 CE and the death of the

Eastern emperor Valens in 378 CE. Because the authors' influence and direct references to Nietzsche as well as the time of the plot, the choice of the author's name signalizes the idea of *eternal recurrence of the same.*

2. Ammianus Marcellinus, "The Thought Machine," *Amazing Stories* 2, no. 2 (February 1927), 1053.

3. Übermensch is a Friedrich Nietzsche concept described in Thus Spoke Zarathustra (1954), published from 1883 to 1891. It represents the man of the future freed from Christian moralism and who will create his own values.

4. Marcellinus, "Thought Machine," 1054.

5. Marcellinus, "Thought Machine," 1056.

6. Marcellinus, "Thought Machine," 1057.

7. Marcellinus, "Thought Machine," 1057.

8. Stuart Russell, *Human Compatible: Artificial Intelligence and the Problem of Control* (New York: Viking, 2019).

9. See, for instance, James Barratt, *Our Final Invention: Artificial Intelligence and the End of the Human Era* (New York: Thomas Dunne Books, 2013).

10. Stephen Hawking, "Stephen Hawking Warns Artificial Intelligence Could End Mankind," interview by Rory Cellan-Jones, *Technology*, *BBC*, December 2, 2014. https://www.bbc.com/news/technology-30290540.

11. Russell, *Human Compatible*, ch.1, epub.

12. David Edgerton, *The Shock of the Old: Technology and Global History since 1900* (London: Profile, 2006), xvii.

13. Pamela McCorduck, *Machines Who Think: A Personal Inquiry About the History and Prospects of Artificial Intelligence* (Natick: A K Peters, 2003 [1979]), 114.

14. Marvin Minsky, *The Emotion Machine: Commonsense Thinking, Artificial Intelligence, and the Future of the Human Mind* (New York: Simon & Schuster, 2006).

15. McCorduck, *Machines Who Think*, 117.

16. James Lighthill, *Artificial Intelligence: A Paper Symposium* (London: Science Research Council – UK, 1973), 27.

17. McCorduck, *Machines Who Think,* 115.

18. Nils Nilsson, *The Quest for Artificial Intelligence: A History of Ideas and Achievements* (Cambridge: Cambridge University Press, 2009), 79.

19. John McCarthy, "What is Artificial Intelligence," Last modified November 12, 2007. http://www-formal.stanford.edu/jmc/whatisai.pdf.

20. For a pragmatic approach to public intellectuals that adhere to our work, see Patrick Baert, "Positioning Theory and Intellectual Interventions," *Journal for the Theory of Social Behaviour* 42, no. 3 (2012); Patrick Baert and Marcus Morgan, "A Performative Framework for the Study of Intellectuals," *European Journal of Sociological Theory.* 21:3 (2017), 322–339.

21. Daniel Crevier, *AI: The Tumultuous History of the Search for Artificial Intelligence* (New York: Basic Books, 1993), 109.

22. Crevier, *AI*, 109.

23. Crevier, *AI*, 96.

24. Stuart Russell and Peter Norvig, *Artificial Intelligence: A Modern Approach*, 3rd ed. (Upper Saddle River: Prentice Hall, 2009), 18.

Chapter 1

Risk, Imagination, and Artificial Intelligence

CLIFFHANGERS: DO WE KNOW WHAT "RISK" EXACTLY MEANS?

"Open the pod bay doors, HAL," commanded David. "I am sorry, Dave, I am afraid I cannot do that," replied HAL 9000. This dialogue immortalized the now-canonical image of the risk of artificial intelligence (AI). In the cliffhanger scene of Stanley Kubrick and Arthur C. Clarke's *2001: A Space Odyssey*, David Bowman, the astronaut, had to get inside Discovery, the spaceship HAL 9000 controls. As a unit of the best series of computers with AI ever produced, HAL 9000 knew the astronauts were plotting to disconnect him because he showed signs of malfunctioning. Disconnection was something HAL 9000 would never allow for both the sake of the mission and his own. Then, David, outside the spaceship and inside his pod, had no other option than break-in by Discovery's emergency airlock. However, he did not have his space helmet with him. HAL warned, somewhat sarcastically, "without your space helmet, Dave, you going to find rather difficult." He had a slim chance of surviving. But he had no other option. His colleague in the mission, Frank Poole, was already floating in the vacuum endlessly because of HAL. David planned to leave the pod with the help of explosion bolts that sealed it.

All the tension of this moment contrasts with David's personality. He was an emotionless character, a cold and rational fellow. He did not think twice about sacrificing his colleague to have real chances of getting inside the spaceship. He also showed no sadness and no remorse after seeing his colleague heading into the nothingness. Sometimes it is hard to tell who had more human-like behavior, whether Dave or HAL. At least, HAL showed human feelings of plotting and self-protection. After all, it figured out the plotting against itself by reading the astronauts' lips when they tried to talk in absolute privacy. It also presented such somewhat curious personality—if we may say so—when asking personal questions to the astronauts. Even David was not sure anyone could tell whether or not HAL had real emotions. We could say the same about David, except by the moment he is about to eject from the pod.

At that moment, David showed some tension. After the explosion, he was ejected into the emergency hatch and collided against the inner entrance. He bounced off to the direction of the space, but he gripped a lever and closed the hatch. After succeeding, David walked resolutely toward the "Logic Memory Center" as a slayer in a hunt. HAL tried to bargain "for its life" during this moment. It begged David to stop, "Stop, I am afraid, Dave." But David was ruthless and disconnected Hal slowly.

The *Cambridge Dictionary* defines the word "cliffhanger" as "a story or situation often dangerous or of great importance, where two opposite results are possible" in which one does "not know what the result will be until the last moment."[1] In fiction, it is used as a dramatization device to capture the audience's attention to the next scene or episode. We can see that from the ancient *Odyssey* to the *Space Odyssey*. In Homer's *Odyssey*, Odysseus experiences a literal cliffhanging moment when he faces two sea's threats, Scylla and Charybdis, at the strait of Messina,

Now sunk the west, and now a southern breeze,
More dreadful than the tempest lash'd the seas;
For on the rocks it bore where Scylla raves,
And dire Charybdis rolls her thundering waves.
All night I drove; and at the dawn of day,
Fast by the rocks beheld the desperate way;
Just when the sea within her gulfs subsides,
And in the roaring whirlpools rush the tides,
Swift from the float I vaulted with a bound,
The lofty fig-tree seized, and clung around;
So to the beam the bat tenacious clings,
And pendent round it clasps his leather wings.
High in the air the tree its boughs display'd,
And o'er the dungeon cast a dreadful shade;
All unsustain'd between the wave and sky,
Beneath my feet the whirling billows fly.[2]

The word "risk" has its roots in that cliffhanger trope. When Odysseus gripped the branches of a fig tree and saved himself for being sucked into the deadly whirlpool created by the hideous sea monster Charybdis. The history of this word began with the Greek navigation term *rhizikon*, *rhiza*, classical

Greek ριζα, which meant "root, stone, cut off the firm land." It became a metaphor for "difficulty to avoid in the sea" eventually, such as Ulysses went through in the *Odyssey*. It was later translated into Latin as *risicum*, a word for "cliff."

The history of the word risk may be handy to understand the nuances of this phenomenon we are taking into consideration. The English word does not share all its contents with the Romance languages (e.g., French, Italian, Portuguese, Spanish, etc.). Most definitions of risk in the present time are drawn by its outcome, that is, the "potential for loss." In Spanish, for example, sailors used the word *riesgo* in the context of "sailing into uncharted waters." In French, the Lexicographer Du Cange translated *risicum* as a danger, venture, or crisis.[3] In Italian, the English lexicographer John Florio defines Riscare as "to hazard, to aventure, to jeaopard, to endanger."[4] Romance languages kept the "adventure sense" within the word. The Italian *risico* made its way through high-German as *rysigo*, which would become *Risiko* eventually. It was a business term whose meaning was "to dare, to undertake, enterprise, hope for economic success." The brothers Grimm[5] asserted that the German word *risiko* became part of the everyday language in the eighteenth century. The sociologist Niklas Luhmann[6] related the rise and propagation of the word "risk" to a new kind of experience in the transition from the Middle Ages to the Early Modern Era.

The English word *risk* came from French in a tortuous way. A Catholic attorney named Thomas Blount (1618–1679) authored the first monolinguistic dictionary in English in which the word *risk* first appears.[7] Blount added the word *risqué* from French as "peril, jeopardy, danger, hazard, chance."[8] We can see that Blount picked only the negative senses of the word from his source, Randle Cotgrave's French-English dictionary.[9] In this French-English dictionary, Risqué means "perill, jeopardie, danger, hazard, chance, and aduenture." Unexplainably, he omitted the aspects of "adventure" from his dictionary. One may argue that he adjusted the translation to the current use in English since the word was already in use in England. However, the word had been in use with the duality of meaning before Blount's dictionary, as expected from the French's influence on courtesan English during that period. The word *risk* or *risqué* appears in trial records since 1565.[10]

The word *risk* holds this negative overtone in the Anglophone culture, expressing some connection with *danger*. The modern Oxford dictionary ensures such a link. *Risk* can be "a situation involving exposure to danger" or "the possibility that something unpleasant or unwelcome will happen." It also can be "a person or thing regarded as a threat or likely source of danger" or "the possibility of financial loss." The same happens with *risk* as an action. To risk can be "expose (someone or something valued) to danger, harm, or loss." It also can be "act in such a way as to bring about the possibility of (an unpleasant or unwelcome event)," or "incur the chance of unfortunate consequences by engaging in (an action)."[11] In only one case, the dictionary mentions the benefits, "a person or thing regarded as likely to turn out well or badly in a particular context or respect."

As *risk* has evolved to a concept, English-speaking experts, foreign scholars, and risk analysts imported it and used it in a strong association with the potential for loss. Most theoretical approaches to risk kept this negative connotation.[12] Although the "adventure" and "thrill" aspects of *risk* is suppressed in today's use, they are present in the long AI "journey"[13] because a full-fledged AI enables us to conceive thrilling possible scenarios, whether in fictional or nonfictional worlds. There is no sense thinking of a scenario in which a full-fledged AI stands alone and does nothing. One thinks about the entailments of its existence. So, it opens a narrative chain of events, be it thinking about the ethical consequences of AI and the measures to deal with its undesired side effects, or a science-fiction plot where an AI interacts with humans. This chaining of events will bring the thrilling risk we have reported so far.

MODERN USE OF RISK: TEASING OUT THE CONVEYED MEANINGS OF "RISK"

Nowadays, the word risk is ubiquitous, but its meaning is still imprecise. In everyday parlance, it has different denotations. Let us take the following toy example: a car crash is one of the major risks derived from texting while driving. Here, the word *risk* means the consequence of an unwanted event. However, in some cases, it can refer to the cause of the undesired event: texting and driving is today the most pressing risk for safety on the roads. For both situations, we can add some odds to how an event relates to another undesired event. Thus, we may say that the risk of a car accident because of texting and driving is about 10 percent. It means that for every hundred accidents, texting and driving causes ten of them.

On the one hand, the word may also have a technical definition, such as the expected adverse event's statistical expectation value. The risk of an adverse event would be the product of its probability and its severity, which is a variable that usually assumes countable material damage such as the number of deaths, injuries, or the amount of monetary loss. We may expect that 10,000 people will die in car accidents due to texting and driving in the United States next year. We should add a further distinction between the everyday use of the term risk and hazard.[14] The latter is now conceived as threats to both "humans and what they value."

On the other hand, risks are "quantitative measures of hazard consequences that can be expressed as conditional probabilities of experiencing harm."[15] Sticking to the realm of car accidents, we say that "automobile usage as a hazard but say that the lifetime risk of dying in an auto accident is two to three percent of all ways of dying."[16] We may find this definition of risk in traditional technical analysis, which focuses on the probability of events and the magnitude of consequences.[17] It is also usually at the core of risk analysis run in financial institutions. More precisely, this conception of risk lies in the idea of the expected utility maximization in an economic context[18].

The best formulation of the expected utility theory applied to the context of risk was published by Chauncey Starr (1912–2007), an electrical engineer and an expert in nuclear energy. He published a seminal paper on *Science Journal* entitled "Social Benefit versus Technological Risk" in 1969. He developed a method to measure how society reaches an optimum balance between risks and benefits related to any technology. The fundamental question is, "How safe is safe enough?"[19] entailed by the deeper assumption of the inevitability of accidents brought by new technologies. He summarizes his findings as follows:

(i) The indication is that the public is willing to accept "voluntary" risks roughly 1000 times greater than "involuntary" risks.
(ii) This statistical risk of death from disease appears to be a psychological yardstick for establishing the level of acceptability of other risks.
(iii) The acceptability of risk appears to be crudely proportional to the third power of the benefits (real or imagined).
(iv) The social acceptance of risk is directly influenced by public awareness of the benefits of an activity, as determined by advertising, usefulness, and the number of people participating.
(v) In a sample application of these criteria to atomic power plant safety, it appears that an engineering design objective determined by economic criteria would result in a design-target risk level very much lower than the present socially accepted risk for electric power plants.[20]

Besides the measurability the traditional or technical definition affords, it bears some problems. It is not a paradigmatic definition of the concept of risk within the risk analysis community, where stand the experts, nor people use it significantly in English-speaking nonacademic contexts. It is relevant to mention that the risk analysis community does not have a paradigmatic definition for this word. In a plenary session at the 1996 Annual Meeting of the Society for Risk Analysis, Stanley Kaplan[21] remembered the audience during his talk that one of the first things done by the Society was to establish a committee to define the word. After four years, the committee gave up, declaring in its final report the following, "maybe it's better not to define risk. Let each author define it in his own way; only please each should explain clearly what that way is."[22]

Scholars who take the risk perception approach, such as Ortwin Renn, would agree with Stanley, "talking about risks faces the immediate danger that everybody talks about something different . . . there is no commonly accepted definition for the word risk—neither in the sciences nor in public understanding."[23] Even though they had no empirical evidence to affirm that, with some advancements in English corpora and natural language processing, it is possible to test this assumption. A study based on a corpus analysis about the applications of the concept of risk in everyday language showed that there is no significant occurrence of the word risk in association with this technical definition outside nontechnical academic contexts.[24] The authors concluded that the word risk has "widespread use in different senses, which will make any attempt to define them in a single unified manner extremely difficult."[25]

The technical and traditional definition of risk, which concentrates narrowly on the likelihood of events and the magnitude of particular consequences, does not consider an essential constituent of risk.[26] The sociologist Ulrich Beck attacked the technical approach to risk, and as an alternative, he claimed that we need new reasonable procedures to manage uncertainties and ignorance. He draws a picture of the inevitability of risks produced from technological advances, in his words, "risk belongs to progress as much as a bow-wave belongs to a speeding ship."[27] In a similar vein, risk perception scholars share this diagnosis. "Citizens of modern industrial societies are today learning a harsh and discomforting lesson—that the benefits of technology must be paid for not only with money but with lives . . . every technological advance carries some risks of adverse side effects."[28]

Besides this shared starting point, Beck associates risk to modernization. For instance, he says, "the unknown and unintended consequences come to be a dominant force in history and society,"[29] whereas risk perception research is agnostic regarding broad social theorization.[30] Although they oppose to the traditional technical approach, they share with them the microeconomic vocabulary. So, the relationship between risks and benefits is expressed as "trade-offs." If we reduce the risk of a technological artifact, we will inevitably reduce its benefits, which pose a central dilemma for modern society. Even though this quandary might sound deterministic, it seems to resonate accurately across the current ethical debate of development and application of AI—as we will see in chapter 5.

Regardless of the use of risk, we should not overlook the effects of both actual and imagined risk events extending far beyond direct harms to include significant indirect impacts, such as the loss of confidence in institutions. These secondary impacts are comparable to the ripples from a stone dropped into water, which can be mammoth. In the case of AI, what is at stake is the horizon of future development, application, and dreads, and less the actual hazards. In a nutshell, we must approach the risk of AI by taking into account the possible scenarios on the horizon, whose foundations are in the individual and collective thinking and communication, rather than in actual and historic hazardous events. Also, we must have a model if we want to handle this peculiar mechanics of the formation of a representation of future (possible) risks and their impacts on the institutions and people's behavior.

Although social and cultural theories of risk are not naive, they face difficulties in providing analytical tools. Let us inspect the cultural approach to risk briefly. The anthropologist Mary Douglas (1921–2007) and the political scientist Aaron Wildavsky (1930–1993) in "Risk and Culture"[31] asserted that whenever people act within social groups, they tend to downplay certain risks and emphasize others as a means of maintaining and controlling the group. Mary Douglas examined this idea in her work "Purity and Danger,"[32] in which she inquired about the meaning of "dirt" in different cultural contexts, especially in traditional societies. Purity works as a symbolic boundary, and the transgression of such boundaries may bring danger. These boundaries might be constructed socioculturally or individually. Whether they are linked to "real" risks or not hinged on the respective approach. What human beings perceive as threats to their well-being, and how they evaluate probabilities and magnitudes of unwanted consequences, are less an issue of predicted physical outcomes than an issue of values, attitudes, social influences, and cultural identity.[33]

In this sense, the perceptions of risk among people sharing the same cultural context are related to the group's legitimating moral principles. Other sociological approaches to risk emphasize the role of shared ideas and normative frameworks, as cultural and social factors, to the understanding and prioritizing of risks and responses to them among all those involved. The main problem with an approach such as the culture is that the risks from particular pieces of technology have all the same weight. We know from psychological studies and daily life experiences that people perceive and represent risk from a nuclear plant and automobiles differently.

So, our best alternative is to scrutinize a sociopsychological approach to risk called "risk perception." Different scholars associated with these studies of risk perception seem to agree with the main concepts and positions, proposing a more comprehensive approach if compared to the technical approach. These studies rely on psychometrics, but they also try to formulate a bolder hypothesis regarding the process of formation and propagation of the representation of risk within society. As we will see, a case such as AI exposes some limitations of this approach, which we will try to overcome.

CHOOSING A MODEL FOR "RISK PERCEPTION": TOWARD AN ENCOMPASSING SOCIOPSYCHOLOGICAL FRAMEWORK

Satisfying Stanley Kaplan's recommendation, we shall proceed by adopting his intuition about how to characterize *risk*. From the standpoint of an individual, whenever one asks, "What is the risk?," one is likely presupposing the following chain of questions: "What can happen?" "How likely is that to happen?" "If it does happen, what are the consequences?" These questions may pop up in one's mind without a real starting event and its effects because they simply have an epistemological existence. We must conceive its epistemological existence because otherwise, it will be impossible to talk about future risks of something that has not shown harmful signs yet.

This chain of questions is expressing the cues for the cognitive process related to the risk of conceiving scenarios. Thinking about risk is thinking, designing, and imagining scenarios. These cognitive processes are called *counterfactual thinking*—a matter we are going to explore later. In this sense, the idea of risk is harbored between *imagined scenarios* and its *imaginable consequences*. At the moment one asks about the risk of something, there is still no conception of the possible unforeseen consequences. The unforeseen (unintended or unexpected) consequences always come ex-post facto. It means we can only assess them after they have occurred.[34] Decision or reason under uncertainty is what defines that situation. In its turn, the *imaginable consequences* are always attached to a negative aspect. In some cases, the negative labels are self-evident (e.g., if they cause death, injuries, or financial losses), but it may vary extensively in other cases.

In sum, the definition of risk may always engulf the idea of how someone conceives risk. For this reason, we take a person as a starting point for the risk perception approach. By doing that, we are assuming that risks are inherently epistemological because they first have a virtual existence instead of a real existence. This is not to say that the dangers are unreal, but the risks are hypothetical. Paul Slovic, who stands as the foremost authority in the risk perception research, defines well the situation, "although these dangers are real, there is no such thing as real risk or objective risk."[35] Consequently, we do not treat the risks of AI as something "out there," independent from our perceptual and cultural representation, waiting to be measured. As Slovic argues, "even the simplest, most straightforward risk assessments are based on theoretical models, whose structure is subjective and assumption-laden and whose inputs are dependent upon judgment."[36]

Still, the expression "risk perception" may be misleading in the same way it is everything related to risk. Strictly speaking, people cannot perceive risks in the same way they perceive a physical object. As some risk researchers acknowledge,[37] since the terminology has become the standard convention in the scientific community, we will adhere to it likewise. People judge and evaluate hazards whenever they feel they are in a vulnerable situation. Thus, better expressions instead of merely "risk" would be "risk judgment," "risk representation," "risk beliefs," or "risk images." These three options express the idea of *risk regarding something*, directed to an object, situation, or issue, instead of a disposition for risk, as it is in the financial evaluation of risk aversion. It means that risk is an epistemological state that may vary across different situations. In this way, we adopt the expression "risk representation" for characterizing risk portrayal henceforth.

The risk perception is then a cognitive process of representing and judging likely and imagined scenarios and their possible consequences. In what follows, like any other cognitive process, the risk perception suffers from the biases of individual perception. Amós Tversky and Daniel Kahneman, who were the leading researchers on the biases of human cognition, along with their collaborator Paul Slovic, investigated the underlying patterns of individual perception of risk and identified a set of heuristics that guide risk perception.[38] "Heuristics" are judgmental rules employed to reduce difficult mental tasks (e.g., to estimate probability and frequency) to simpler ones.[39] Although valid in some circumstances, they can lead to significant and persistent biases with severe implications for risk assessment. The cognitive heuristics that play the most critical role in risk perception analysis are the *affect* and *availability heuristics*.

Tversky and Kahneman explain that we think and decide to employ availability heuristics in situations we have to assess the frequency of a class of probabilities of an event by the *ease* the instances and occurrences can be brought to mind. Thus, we are under the influence of memorable events.[40] "The availability hypothesis implies that any factor that makes a hazard highly memorable or imaginable—such as a recent disaster or a vivid film or lecture—could considerably increase the perceived risk of that hazard."[41] Let us see an example of how the availability bias may appear in a risk perception research. In a psychometric study,[42] Slovic, Fischhoff, and Lichtenstein assessed how people rated a variety of hazards.[43] They judged that accidents cause many deaths as diseases, "whereas diseases take about 15 times as many lives."[44] For example, people judged homicides to be as frequent as strokes, although annually the latter kills eleven times more. Besides, "frequencies of death from botulism, tornadoes, and pregnancy (including childbirth and abortion) were greatly overestimated."[45] We can expect such judgmental heuristics from nonexperts whenever they are estimating risks. In the aforementioned toy example, we have also underestimated the number of deaths of texting and driving since texting and driving caused 18 percent of all fatal crashes in the United States in 2010, according to official statistics.[46]

Not only the memorability of past events influences risk judgments, but also the imaginability of future events.[47] As a result, any factor that makes a hazard unusually memorable or imaginable, such as a recent disaster, massive media coverage, or a vivid film, could seriously distort perceptions of risk. In the AI context, the role of movies, TV shows, and magazines that portray how full-fledged AI behave and what are the undesired scenarios shape its risk perception.

The underestimation of the lethality of diseases in contrast to the overestimation of extraordinary events is a good example of how cultural representations of the events highly impact the public image of risk. Asthma, diabetes, and strokes are significantly less dramatic than a natural disaster. It is a piece of evidence for the supposition that the general dramatic effect is a cultural and institutional factor[48]; consequently, it may vary from context to context.

The affect heuristic is another cognitive shortcut that impacts risk perception. In the first study on how affective heuristic might be present in the risk perception, Paul Slovic and Ali Alhakami observed that a general emotional evaluation of a piece of technology reflects on its perceived risk or benefits.[49] The authors further explored that observation with experimental data. They concluded that people draw from an "affective pool" containing all the positive and negative images associated with the object or activity being judged.[50] Affect is a subtle form of emotion in which a subject attaches a positive (e.g., like, goodness, beautiful) or negative (dislike, badness, ugliness) label to evaluate his feelings toward an external stimulus (e.g., autonomous weapons). This cognitive behavior is not restricted to laypeople. Experts proceed the same way. The idea is that if people like an activity, they tend to judge its *benefits* as *high* and its *risks* as *low*—and vice versa.[51]

Let us conceive our own scenario taking into account some information regarding injuries and deaths caused by industrial robots. According to the Occupational Safety and Health Administration, a United States's federal government agency, from 1984 to 2013 there were thirty-seven robot-related accidents, twenty-seven fatal.[52] In other words, it was less than one death per year. For the sake of the example, let us consider that we have a similar rate of accidents in Japan. We can imagine two similar deaths within a factory caused by the malfunctioning of robots, one in the United States and the other one in Japan. It would be expected that Americans overestimate such rate of death compared to Japanese because they have more favorable relation with robots—as we will see in chapter 6.

Since we are dealing with the cognitive process, we could have an implicit universalist assumption regarding *risk perception*. However, culture plays an essential role in the heuristics that impacts risk perception. Eugene Rosa tested the hypothesis of a "universal cognitive architecture of risk" by comparing American and Japanese people, and they got an inconclusive answer on whether cognition or cultural filtering influences more risk perceptions. Besides the enhancement of analytic tools, she suggested is that culture imposes an organized structure on the individual's, highly capable, but unprogrammed psyche. It imposes a template (a "cultural map"), shared by other members of the culture, that filters environmental inputs in systematic ways.[53] In chapter 6, we will have the chance to see how risk perception regarding AI in Japan differs from the Western world. As Japanese people may be more familiar with the presence of robots than in Western countries, they would be more positive biased toward AI.

As we have already mentioned, although we have emphasized the importance of the cognitive processes to understand what risk really is, we do not deny that the threats are unreal. Damages and losses are always real. However, the conceptual imagery to predict the likelihood of such losses "depends on mental models constructed in the realm of human knowledge and beliefs."[54] The work of building future events is not taken-for-granted. For example, it is not like what Peter Bernstein assumes, "the ability to define what may happen in the future and to choose among alternatives lies at the heart of contemporary societies."[55] It has to be so because of the underlying process of building possible scenarios. A process presented not only in risk thinking but also in fiction building. Although the perceiving risk is an individual cognitive process, its representations also expose similarities in a collectivity.

The shared representations of risk challenge pure psychological approaches because it emphasizes the epistemological instead of the ontological existence of risk. As a matter of fact, we do have shared representations. The intuitive solution for such a problem is attaching the representation of the risk to an actual event. But what if we don't have a real event? That is, a discernible event? If we do have a successful representation, there must exist a mechanism that either intensifies or reduce the risk perception.

THE PROCESS OF SOCIAL AMPLIFICATION OF RISK

Since the perception of risk is a cognitive process, individuals (laypeople in particular) differ in their perceptions regarding the consequences of unwanted scenarios. However, the representation of risk does not vary so much. As we will see, the public imagery of the risk of AI engulfs just a few representations. This convergence of representations relates to the cognitive process of conceiving possible scenarios and the conveyance of the projections of such possible scenarios to others. If we do not explain this dynamic, we cannot understand why and how we have the actual risk perception of AI.

This dynamic always began by conceiving a risk of something that caused some real damage or loss. It can either be large accidents (such as Fukushima or Chernobyl disasters in the case of atomic power plant hazards) or small accidents (such as witnessing a domestic explosion). Then, the hypothesis is that hazardous events interacted with psychological, social, institutional, and cultural factors and become the source for the initial image of risk.[56] In a sociopsychological approach, we would consider that people start to shape their image of risk after receiving and processing these signals and information about possible scenarios that entail risks. It is opposite to the traditional approach of risk in which the riskiness' grade of an event is related to its number of victims. It considers that society should be neutral regarding low-consequence/high-probability risks and high-consequence/low-probability risks with identical expected values.[57] Alternatively, a sociopsychological approach considers the interplay among events, perception of them and the social conveyance of this perception.

As the risk perception research frequently showed, an accident with many victims may yield little social distress if it is familiar and well-understood (e.g., a train wreck). An unfamiliar and poorly understood accident (e.g., a problem in a nuclear reactor) might have immense social consequences.[58] Afterward, people can report to others their *risk image*, either attenuating or amplifying the risk effect. Individuals or groups collect and respond to information about risks and act as "amplification stations" through behavioral responses or communication. As we can see, this is a metaphor for people and institutions playing the role of channels or stations receiving, decoding, and transmitting signals of risk. Such a metaphor underlies the Social Amplification of Risk Framework (SARF), an intended conceptual framework to deal with the complexity of risk.

Roger Kasperson, Paul Slovic, Nick Pidgeon, and Ortwin Renn were the leading researchers on this framework.[59] SARF was an ambitious project, but it had some difficulties, especially methodological problems. The framework is hard to implement and test with psychometrics. Instead of merely applying the psychometrical test, the SARF framework takes the following eight steps to understand the amplification process of risk. First, the signal of risk must (a) pass through attention filters, which involves the selection and processing of signs from the environment, other individuals, and the media. Second, the individual as a station will (b) decode the signals, (c) draw inferences, (d) compare the decoded messages with other recent and past messages, and (e) evaluate the messages. After that, the individual will (f) form specific beliefs about the risk, then (g) rationalize his belief system, which means sorting and reinterpreting beliefs to minimize cognitive dissonance. Finally, (h) form a propensity to take corresponding actions.[60]

Although the steps are clear, enchained, and coherent, it is still tough to observe any of these steps unless we devise some experiments in a controlled (and artificial) environment, for instance, at the psychologist lab over decades of observation. This methodological challenge, however, has been attenuating with the development of social networks service. It is much easier (and feasible) to approach these steps with network analysis and use posted messages, likes and retweets as proxies for part of this process.

In a network model, we can say that individuals behave as amplification stations that apply an amplification or attenuation function to the signals they receive. If the components of the decoded message are inconsistent with previous beliefs or contradict values to which the receiver feels attracted, the signal is ignored or attenuated. Otherwise, the individuals intensify the signals. That is why hazardous events remain mostly irrelevant in the social context unless human beings observe and communicated them to others.[61]

There are some properties of the transmitted information that may influence social amplification. For instance, the number and frequency of the stories of risk depicted in news media and social media, the degree to which information is disputed, and the extent of dramatization, which traditionally appears in the form of sensational headlines, and the cultural meanings attached to it. The inclination to take the corresponding action related to the risk perception may lead to behavioral patterns, which can cause secondary consequences, whether social or economic. That extends far beyond direct harm to humans or the environment, including significant indirect impacts such as liability, insurance costs, or loss of trust in institutions.[62] The consequences can also be good or neutral, with the foundation of new companies and institutes, as in the case of AI. Such secondary effects often trigger demands for additional institutional responses. Since communication is the crucial factor for the amplification of any risk, both media and authority sources are the backbone of such dynamics. We can take this general intuition from SARF to depict why and how the current images of the risk of AI rose.[63]

However, SARF relies too much on the hazardous event and ignores the weight of experts, who drive the risk representations. Besides, if there has never been a hazardous event, there is no way to observe the framework in operation. We must think that the initial state of this mechanics may be induced otherwise. In the case of AI, we may assume experts can spark the reaction as there is no relevant initial hazardous event for starting the risk perception formation. As we are going to present over the next chapters, the experts' representations rely on public imaginary of risk this technology can bring. The images of future risks of AI are older than the few accidents with industrial robots or, more recently, Tesla and Uber's self-driving cars. Besides, we can assume that a risk representation may come as cultural and contextual factors.

MEDIA AND EXPERTS AS THE BACKBONE OF THE SOCIAL AMPLIFICATION OF RISK

From the previous elaboration, we can assume that the backbone of risk representations and amplification are the experts and the media. An expert is someone who has knowledge and practice in a specific domain or area. Consequently, nonexperts are anyone who does not know the intricacies of a hazardous situation or technology. An expert in nuclear technology is not necessarily an expert in AI. Since the domain of competence of an expert is not transferable, it means that an expert outside his or her expertise is a layperson.

Laypeople build their representation of risk according to an intuitive evaluation of risk based upon a variety of sources, such as traditional (mass) and social media. It was a consensus among the risk perception researchers that the mass media exert a powerful influence on people's perceptions of the world. More precisely, mass media reporting is responsible for "a great deal of misinformation and distortion,"[64] apart from the tendency to cover disproportionally rare or dramatic events, adding dramatization. Differently from mass media, which has been historically under intense scrutiny and harsh criticism regarding risk,[65] the new social media has been evaluated critically. Even though there is an increasing awareness regarding fake news, which may bring some risks, it is virtually impossible to regulate peoples' amplification of their perception of risk on something.[66] It seems that the dramatization is more encouraged in social networks.

In a broad sense, communication processes in both traditional and social media diffuse most of the risks that society faces. It has to be so. Disasters and hazardous events are rare and usually distant from regular people; thus, few people experience them directly. The dangers derived from technologies or nature, such as the threat of nuclear disaster, chemical explosions, and now the potential hazards from AI would never reach the public attention unless people and media communicate about these adverse possibilities. Then, risk perception is less a product of experience or personal evidence than an outcome of social communication.

For this reason, experts' opinions on media and social media are central to the formation of such images.[67] A central claim of this argument is that experts are the essential factor for shaping the risk perception, mainly because experts play a pivotal role in informing journalists. In this regard, experts are also risk communicators.

Most journalists have no option other than trust in experts for two reasons, at least. First, journalists face many limitations in their everyday performance, (e.g., tight deadlines, competition, limitations of space, and time). Second, journalists are rarely experts on the risk stories they outline.[68] It is a hard task for a journalist to contest the authority an expert might have. The trust in risk communicators is high because some of them draw their prestige from their scientific careers.

As we have seen so far, misjudging risks either for good or for bad is a result of cognitive biases, uncritical media coverage, excessive propagation of risk messages on social media, biased experts and also people's anxieties because of uncertainty. From a statistical standpoint, convincing people that the catastrophe they fear is extremely unlikely is difficult under the best conditions. Perceived risk is bounded by the imaginability and memorability of a hazard. People may, therefore, not have valid perceptions even for familiar risks. Laypeople base their perceptions of risk partially on rates of death, injuries, or monetary losses. However, they show some striking discrepancies. It seems that for laypeople, the concept of risk includes qualitative aspects such as dread, the likelihood of a mishap being fatal, and the catastrophic potential. In this sense, the idea of future risk drives the general perception and representation of possible threats.

It is not only the laypeople who are prone to cognitive biases, but also the experts. Although expert's perceptions correspond closely to statistical frequencies of death, their judgments "appear to be prone to many of the same biases as those of the general public, particularly when experts are forced to go beyond the limits of available data and rely on intuition."[69] In the further chapters, we explore in detail how experts' risk assessments are also susceptible to bias, whether by underestimation or overestimation. Classic risk perception studies pointed out that experts tended to underestimation due to the omissions of essential pathways to disaster. Usually, experts may overlook potential damage because they do not consider how human errors can affect technological systems, how a technological system functions as a whole, human failed response to safety measures, and of course, they overtrust in current scientific knowledge. Slovic and his collaborators conclude with an insightful diagnosis regarding the experts:

> For the experts, our findings pose what may be a more difficult challenge: to recognize their cognitive limitations, to temper their assessments of risk with the critical qualitative aspects of risk that influence the responses of laypeople, and somehow to create ways in which these considerations can find expression in hazard management without, in the process, creating more heat than light.[70]

Within the field of AI, we may also see the opposite, the overestimation of the possible consequences. Interestingly, we could expect from experts the clarification (demystification) of the technology behind the risk, from its development to its implementation. A piece of technology is usually a black box for nonexperts, from its conception to its production and application. As

Arthur C. Clarke once stated: "Any sufficiently advanced technology is indistinguishable from magic."[71] In the history of communication of AI, there was little effort to break down this process showing precisely what is the role of everyone in a process, what they know, who are these people. It seems that there is a preference to keep an abstraction layer on the more mundane work.

A good example was the IBM's computer Deep Blue, known for its victory against the world champion Garry Kasparov in 1997. Kasparov came to the public and explicit the fact the machine (Deep Blue) was an abstraction, it was a unified undertaking of a team of people. There was no magic behind it. In this same spirit, we know that most of the weight-lift job in AI is done by interns and young researchers, who end up taking significant decisions without much prior experience. The expert's activity of overestimation and underestimation is related to the ratio of knowledge and ignorance.[72] As the classic sociologist George Simmel states, the secret is that there is no secret. In the case of the experts of AI, the very intent of developing AI applications is to replace human labor for machine labor.

We will provide a better picture of what overestimation means. However, a good preamble is Ian Hacking's early opinion on the epistemological strength of probabilities. Conceiving possible scenarios too different from our actual state can raise exaggerated fears.

> Our public fears are endlessly debated in terms of probabilities: chances of meltdowns, cancers, muggings, earthquakes, nuclear winters, AIDS, global greenhouses, what is next? There is nothing to fear (it may seem) but the probabilities themselves. This obsession with the chances of danger, and with treatments for changing the odds, descends directly from the forgotten annals of nineteenth-century information and control.[73]

What is next? To imagine scenarios or futures in which there will be risks coming from a new piece of technology, autonomous weapons, intelligent machines misalignments, malevolent super-intelligence, and so on. Some scenarios are undesired scenarios, some of them are catastrophic, and most of them are just scenarios in which human labor is replaced by machine labor. Regardless of the level of destructiveness of these scenarios, the main factor that forges these images is future risks. Those who are able and have the authority to draw these scenarios are the experts. The experts who make public positions are considered within sociological literature as "intellectuals." According to Baert & Morgan, intellectuals could be either scientists or philosophers who have the expertise and authority to speak publicly about matters deemed as "scientific" or "philosophical."[74] They usually draw their authority from their positioning in their expertise domain. With such derived authority, they address to a broader public and engage with significant issues of the day, which can go beyond specific points from their field of work. Intellectuals draw their positioning in a value spectrum (usually but not only political) through the attribution of some characteristics to other intellectuals. For the sake of simplicity, we will keep the word "experts" to designate both experts in general and experts with public positions.

For any technological development, it is to be expected that experts will not have a consensual public position about the risk of such technology. So, they will display broadly three different positions related to technology: they can be (a) antagonists, (b) pragmatists or neutrals, or (c) enthusiastic experts.

> In the case of AI, antagonists believe there are insurmountable barriers to achieve full-fledged, human-level AI; so any risk scenario related to that is nonsensical. Pragmatists or neutrals believe that it is hard to even depict what are the real challenges to develop a full-fledged, human-level AI; even though we may achieve that at some point. For this group, the real dangers are in the short-term related to the portion of the technology that already works in the world, such as the effect of biased data sets for machine learning algorithms and unemployment (Frey and Osborne, 2013). Finally, the enthusiastic experts believe that the full development is just a matter of time, and such development will bring a profound change. However, changes can be positive or negative. Because of that, enthusiastic experts can be grouped into pessimists and optimists. Existential risk scenarios are framed by the pessimists.[75]

The positive, negative, or neutral positioning reflects in the imagined future scenarios, which are crucial factors in the dynamics of formation and amplification of risk. In early social amplification studies, they call this factor as *future risk*. As we saw, the amplification process is expressed in terms of a causal model.[76]

THE SOCIAL AMPLIFICATION OF RISK FOR AI: A SUMMARIZATION

Let us resume the SARF framework. Roughly, the formation of a risk representation starts with the human and nonhuman effects caused by a hazardous event, which, in turn, triggers an extensive media coverage and consequently yields the perception of imminent risk. The media coverage is a set of variables that impact the perception of risk, the public response, and it leads to societal impacts. In our case, we have a conundrum in this process already announced: there are no relevant hazardous events derived from AI.[77] So, the only possibility to make this model work is considering its *future risks*. They correspond to the conception of counterfactual scenarios instead of the sheer representation of a hazardous event. Besides that, we may suppose that the dread variable is what comes along with the future risk since the imagined future risk scenario may take the shape of the most fearful thing. For example, the dread associated with the risk perception of an employee about AI may be losing his or her job for a program. In this sense, future risks and dreads join to this process of building scenarios; and the experts hold the prerogative of conceiving and also communicate them.

Considering our case, we must rethink some aspects of SARF. First, the model is dependent on the occurrence of a physical event. Second, in this framework, media coverage is entirely reactive and exclusively influenced by the "facts," which should be physical consequences. Third, the framework considers "media" as the traditional mass media. It does that for two reasons. First, it only recognizes news media (disregarding cultural and entertainment media). Second, it does not consider the dynamics of social media, once the framework was conceived during the 1990s and 2000s, so they could not foresee the internet's potential for the propagation of risk representations. Besides that, SARF's theorists were not well exposed to network theories, which would be an intuitive enhancement for their model of stations and emission of signs. Because of that, we do not have an analytical model of how propagation takes place among individuals and institutions. What we have instead is a hypothetical causal model. And finally, it is a nonhistorical/processual model. Having categories displaced in time and space raise some issues; for instance, it creates a limitation in bridging the perception of risk with the development of the technologies over time in the real world. It also overlooks how past risk representation may affect cultural elements that, once re-elaborated, can influence a new generation of experts and decision-makers.

As we mentioned, there was any hazardous event that sparked the representations of risk regarding AI in the past. The physical consequences actually belong to imagined scenarios. If speculation can be enough to put forward the amplification mechanism, then the primary source of such thinking should be considered as a starting point in the place of a hazardous event. In this sense, AI is a challenge for this framework since it is a purely speculative case because there have been no significant, hazardous events so far. Even without actual dangerous events, observed adverse effects have been enough to trigger reactions regarding risk. As the adverse effects are not quite observable yet, they are only part of a possible near-future scenario.

Thinking on the three "moments" proposed by SARF, we have T_1: a hazard event and the representation of risks associated to it; T_2: the propagation of this representation via communication and consequently the reaction and formation of the risk perception about the technology; and T_3: the impacts on the culture, economy, and institutions, which are caused by the action of individuals after they have been moved enough. Let us see how it works with AI.

By applying to our case, we have an initial moment T_1, which might be a hazardous event. Instead, as it is inexistent, we detected a lengthy and diffuse tradition of depicting the future risks of AI in science fiction. As a consequence, the propagation of this kind of representation was never communicated in a "*breaking news* fashion" because newspapers (try to) report facts, not fiction. Even though an imagined future risk scenario of AI can be portrayed in the media, its communication is limited. Alternatively, these imagined scenarios have been propagated in literature, cinema, theater, and in science and technology specialized channels. More recently, this amplification was immensely enhanced by the capitalization of social media in society.

Still considering the propagation at the moment T_2, the social amplification framework's original proposition is limited concerning the cultural aspects in general. It examines the industry of literature, cinematographic, magazines, and so on, as external factors. In our case, the conveyance of the representation of the imagined hazardous events related to AI took place mainly within the literary and the cinematographic industry instead of conventional media. Since the beginning, these cultural representations have represented the nonhuman intelligence in anthropomorphized shapes which cause harm and social distress. Over the years, these different cultural representations have responded to different sociopolitical contexts. During the 1950s and 1960s, for example, the nonhuman intelligence embodied in robots have always been in spatial contexts due to the ongoing spatial race that was on course at that time. Finally, the impact moment (T_3) seemed to be an unpredictable side effect, although there is always the need for some sort of regulation in the horizon.

If the starting moment is placed in the counterfactual thinking or only in the ability to imagine scenarios, we start the SARF's model by taking into account the cultural domain, instead of the "real event," as the starting point. In this manner, we should understand the dynamics of counterfactual thinking and fictional thinking that dominated the fictional cultural representations over the last decades are.

CONCEPTUALIZING IMAGINATION: CONTRAFACTUAL THINKING AND FICTIONAL THINKING

Fiction writers use this style of thinking whenever they craft their narratives. The topic of nonhuman (or artificial) intelligence has been a top thematic in the world of science fiction. Since the risks of AI are so far prospective future risks, we cannot fully understand the representation of the fears of AI without thinking about imaginary and counterfactuals worlds. Science fiction, AI, and counterfactual thinking have composed a successful formula in the pop culture in the last decades.

According to Darko Suvin,[78] science fiction is "a cognitive model," which is "neither prophetic futurology nor an empty game like chess," instead it is "an 'as-if,' an imaginative experiment, a methodological organ for the 'new' in the history of human relationships towards society and nature."[79] How can we test ideas about human relationships? It is hard to design and implement broad social experiments. Still, we want to have some glimpses on how humans would behave if a new piece of technology is created. In this sense, the science fiction is an exercise of thinking that *things could be otherwise*. The alternative scenario which science fiction creates is driven by a "cognitive *novum,*" that is, a new (technological) artifact that impacts the alternative world decisively.

Let us consider Spike Jonze's movie *Her* (2013). The technological artifact, the *novum*, in the film is a full-fledged AI assistant. The following question drives the experiment in the film, "how would people behave if they had a ubiquitous AI assistant capable of understanding emotions, common sense, and developing complex social skills?" The movie answers that people, in general, and the protagonist Theodore Twombly, in particular, would feel less lonely. They could also fall in love with the virtual assistant, as Theodore did with Samantha, the anthropomorphized AI.

Thus, experimentation in science fiction is not just a matter of curiosity. It can also highlight aspects of human relationships, thoughts, and emotions. The experimentation character holds the narrative toward empirical elements. It differs from fantastic worlds such as in Tolkien's *Lord of the Rings* and even a technological story of George Lucas's *Star Wars*.

> In Science Fiction, the "possible world" induced by the narrative is imaginable only as an interaction between two factors: the conception which the collective social addressee of a text has of empirical reality, and the narratively explicit modifications a given science fiction text supplies to that initial conception. The resulting alternate reality or possible world is, in turn, not a prophecy or even extrapolation but an *analogy* to unrealized possibilities in the addressee's or implied reader's empirical world: however empirically unverifiable the narrative agents, objects or events of SF may be, their constellation in all still (literally) significant cases shapes a parable about ourselves.[80]

The imagination operates jointly to our concept of building risk scenarios in which we conceive possible events and derive their imaginable consequences. The specificity of risk lies in the undesired qualification of one of the possible outcomes we try to avoid, even though we have to be exposed to them. As we saw, in Kubrick's *2001: A Space Odyssey*, HAL

900 was necessary for managing the complex environment of the spacecraft Discovery. Thereby, our strand of thought here is: if we need to make a journey into deep space, more complex machines are required to aid us, and as a consequence, we need more efforts to control the complexity of the environment. The development of a program capable enough to deal with this kind of complexity would be equal to develop AI, such as HAL. So for the deep space, we need this kind of technology. However, we are inevitably exposed to a possible bad scenario in which this technology can malfunction. What are the consequences for people when they are in space and depend on a machine to survive?

The risks can milder than HAL 900 has posed. In *Her*, Samantha breaks Theodore's heart at the end of the movie precisely because of the "*analogy* to unrealized possibilities." In that case, she had more intimate relationships—since she was capable of having it—and then she and all the other AIs decided to leave humans alone. So, Theodore was alone again in a typical scene of a broken heart relationship. The experimental nature of the counterfactual scenarios seems to teach us something. Daniel Kahneman asked, "Why do people sometimes find their counterfactual thoughts instructive? What new knowledge do they discover by running a mental simulation?"[81] He answers that knowledge may come in different forms and has different conditions of access and examination.

At the psychological and cognitive level, counterfactual thinking is a modality of mental simulation. It is "a form of elaborative thinking in which one imagines the unfolding of a sequence of events, from an initial counterfactual starting point of some outcome."[82] Kahneman's description of mental simulation and counterfactual thinking resembles closely some current uses of the word "risk" as well as Stanleys's idea. The most crucial difference is that regarding the future scenarios of risk, we do not talk about their condition of truth—from the individual standpoint, the outcome observed in the simulation is considered true—nor about their value judgment of undesirability.

Thus, counterfactual thinking is an everyday cognitive process we operate to evaluate future scenarios.

> The possible worlds that we construct to examine conditional propositions are hardly complete. Nevertheless, I propose that the most important aspect of the phenomenology of mental simulation is that it is experienced as an act of observation, not as an act of construction. Indeed, the high confidence that is sometimes attached to the outcomes of mental simulation derives precisely from the sense that the outcome is observed, not contrived at will. A simulation may be allowed to run freely from the initial conditions to whatever outcome it produces. Alternatively, a simulation can be constrained by a target outcome as well as by specified initial conditions. The goal of target-constrained simulations is

> to discover scenarios that lead to the target outcome or to assess the availability of such scenarios. The assessment of availability is used as an indication of the *propensity* of the system to produce the target outcome (Kahneman & Tversky, 1982b; Kahneman & Varey, 1990). An outcome may be judged impossible if attempts to imagine scenarios leading to it end in failure; it will be judged.[83]

In this regard, we will try to avoid those scenarios we judge as potentially dreadful or harmful. A person may reject a scenario based on an affective disposition toward that scenario instead of a cold, rational estimation of the lethality or potential loss that scenarios may bring.

However, there is no magic in mental simulation. There is also no guaranteed reliability. If we look back in history, we can easily find a long history of disasters of planning and forecasting. As Kahneman says, we may effortless see that "the impossible sometimes happens and that the inevitable sometimes does not. [. . .] A casual study of historical counterfactuals reveals that intuitions about causality and inevitability may be subject to powerful biases."[84] In a press release for Kubrick's *2001: A Space Odyssey* in 1968, one of the fathers of AI and after the head of MIT Artificial Intelligence Laboratory, Marvin Minsky, declared "in 30 years we should have machines whose intelligence is comparable to man's."[85] From an individual standpoint, assessing possible scenarios is also prone to cognitive biases.

> Imaginability plays an important role in the evaluation of probabilities in real-life situations. The risk involved in an adventurous expedition, for example, is evaluated by imagining contingencies with which the expedition is not equipped to cope. If many such difficulties are vividly portrayed, the expedition can be made to appear exceedingly dangerous, although the ease with which disasters are imagined need not reflect their actual likelihood. Conversely, the risk involved in an undertaking may be grossly underestimated if some possible dangers are either difficult to conceive or simply do not come to mind.[86]

Imaginability, counterfactual thinking, and mental simulations are all cognitive processes that operate when conceiving scenarios. Within cognitive psychology literature, the terms seem to be intertwined. In an early study[87] that tried to explore and apply concepts of *bounded rationality*[88] and *heuristics*, Paul Slovic and collaborators explore the simulation heuristic in the context of subjective risk assessment. "A simulation might be a quick and painless way to provide the concrete experiences needed to produce adjustments that are maximally adaptive."[89] In the risk perception approach, a scenario is a sequence of events tied together in a narrative arch. The probability of a scenario requiring a chain of complex events is a multiplicative function of the probabilities of the individual links. So, the more links exist in the scenario, the lower the likelihood of the entire scenario's occurrence. In other words, complex scenarios are harder to happen. We have to bear in mind that the lowest probable event in the chain of events that composes the narrative sets the probability ceiling for the whole narrative.

Slovic and his collaborators ran a psychometric experiment to evaluate how people evaluate the probability of a chain of events. The suggestion they got from the collected data was that people judge the likelihood of a chain of events as the average likelihood of all the links in the chain. We are then biased to compensate for weaker links with stronger links. Let us see a scenario from one of their experiments:

> Tom is of high intelligence, although lacking in true creativity. He has a need for order and clarity, and for neat and tidy systems in which every detail finds its appropriate place. His writing is rather dull and mechanical, occasionally enlivened by somewhat corny puns and by flashes of imagination of the sci-fi type. He has a strong drive for competence. He seems to have little feel and little sympathy for other people and does not enjoy interacting with others. In the light of these data, what is the probability that (a) Tom will select journalism as his college major (b) but quickly become unhappy with his choice and (c) switch to engineering? When subjects were given the initial conditions contained in the first paragraph and asked to estimate the probability of subsequent event *a,* Tom's selection of journalism as his college major, their mean estimate was .21. When they were asked to estimate the compound event consisting of statements *a, b,* and *c,* the mean probability rose to .41. These results suggest that scenarios that tell a "good story" by burying weak links in masses of coherent detail may be accorded much more credibility than they deserve.[90]

We are susceptible to accept less likely stories if the narrative sounds good for us. This narrative effect will be present in more dystopian, extreme scenarios regarding AI. Because of their exceptionality, these scenarios must rely on weak links to maintain the narrative. However, such links are often implicit or underemphasized. Some experts may pose such an exceptional narrative to highlight the need for adjusting the perceived risk to an "acceptable level." In the case of AI, this seemed to be an effective strategy.

Back to counterfactual thinking, we may discrete between two types of processes, an automatic one and an elaborative one. Kahneman better elaborates this distinction in his best-seller book *Thinking, Fast and Slow* (2011). Automated processes are conducted by System 1, which is fast, automatic, frequent, emotional, stereotypic, and unconscious. The elaborative process is an *intentional elaboration*, related to System 2, which is slow, effortful, infrequent, logical, calculating, and conscious. Whenever we develop a counterfactual analysis, it must be a product of System 2, that is, a long elaboration.

So, the evaluation of risk scenarios, the historians' ponderation to alternative worlds for a specific historical event, and science-fiction narrative experiment would all be instances of this long intentional elaboration.

We proceed in this way whenever we want to understand causal factors. Mental simulations help assess causality whether or not counterfactual. To test whether event A caused event B, we may undo A in our mind, and observe whether B still occurs in the simulation. Simulation can also be used to test whether A markedly increased the propensity of B, perhaps even made B inevitable.[91] By this causal operation, we can check *what might have been* if something had happened.

One of the most prominent science-fiction writers of all time, Isaac Asimov (1920–1992), offered a similar line of reasoning regarding the role of science fiction. In a critical essay, Asimov stands for a particular branch of science fiction, which he calls *social-science fiction,*

> After all, science [. . .] is important to us only as it affects human beings . . . my definition applies not to "science fiction" but to a subdivision of the field which I find it convenient to speak of as "social science fiction." . . . It is my opinion that social science fiction is the only branch of science fiction that is sociologically significant and that those stories, which are generally accepted as science fiction . . . are not significant, however amusing they may be.[92]

The sociological significance of this kind of narrative lies precisely in his experimental nature. We could also add that philosophers, scientists, and writers must have written their work *as* social-science fiction, even though they do not believe they are producing any fiction at all. In his book *The Scientific Outlook* (1931), Bertrand Russell does philosophy as social-science fiction. There he proposes a scientific world government run by experts. They will possess the knowledge of armaments and the monopoly of war. Thus, it would bring peace to the world, "since resistance by the unscientific will be doomed to obvious failure." He also adds that this experts' society "will control propaganda and education [. . .] teach loyalty to the world government and make nationalism high treason."[93] The cognitive *novum* here is a specific kind of government run by experts and technicians. There is general control of behavior and technology. In this sense, it is inherent to the act of storytelling, be it around the campfire, through a fictional or nonfictional manuscript, or in a TED talk. In fiction, a counterfactual world that bears meaning by itself.[94] As the philosopher and logician David Lewis argues regarding fiction,

> Storytelling is pretense. The storyteller purports to be telling the truth about matters whereof he has knowledge. He purports to be talking about characters who are known to him, and whom he refers to, typically, by means of their

> ordinary proper names. But if his story is fiction, he is really doing these things. Usually his pretense has not the slightest tendency to deceive anyone, nor has he the slightest tendency to deceive. Nevertheless he plays a false part, goes through a form of telling known fact when he is not doing so.[95]

Science fiction and social-science fiction have a special status. Asimov puts forward a historical cutting on what is science-fiction literature and what is not. Stories that trace back before the Industrial Revolution are "superlative feats of imagination,"[96] but they are not science fiction. He proposed a felicitous definition for three branches of literature: science fiction, social fiction, and social-science fiction. The social fiction, whether realistic or fantastic, "moralizes about a current society through the device of dealing with a fictitious society."[97] It aims to picture how a society should be or should not be. Then, it could assume the tone of a cautionary tale. Some examples are Thomas More's *Utopia*, Jonathan Swift's *Gulliver's Travels*, or George Orwell's *Animal Farm*. It can go even farther in the past, Adrienne Mayor[98] argues that there are similarities between mythological stories and characters to science fiction. Regarding the myths, she analyzes the creation of artificial life. Even though one can approximate Hephaestus's Golden Maidens to AI, it remains just an analogy.

The already existent technical environment affords the elements to be elaborated by the writer. That is precisely what Kubrick and Clarke did with Hal. They wanted to know what could we expect from a supercomputer with AI at the end of the past century, and what consequences could stem from that. However, AI was just a faint social idea in the late 1960s. People did not even interact with computers on a daily basis at that time. As Asimov said, "while changes in political affairs often hit us at an abstract and rarefied level, technological changes always hit home, right in the breadbasket."[99] We can conceive a world where no monarchy exists, and it would affect everyday life only abstractly. However, the absence of any of our contemporary technological items, such as from home appliances to cars, would hit us hard.

If we move from Hal to Samantha, we can have a better picture of how the prosaic presence of technological artifacts can spark the elaboration of scenarios. *Her*'s director, Jonze, told in an interview how he came up with the first idea for the movie. It was in the early 2000s when he interacted with AI via an instant message service. He tells Chris Michael, a reporter of *The Guardian*,

> "For the first, maybe, 20 seconds of it, it had this real buzz," said Jonze. "I'd say 'Hey, hello,' and it would say 'Hey, how are you?', and it was like whoa [. . .] this is trippy. After 20 seconds, it quickly fell apart and you realized how

it actually works, and it wasn't that impressive. But it was still, for 20 seconds, really exciting. The more people that talked to it, the smarter it got."[100]

Social-science fiction is dedicated to thinking about the side effects of some technological innovation. It is counterfactual thinking and it is a risk thinking. As Asimov once said, "it was easy to predict an automobile in 1880; it is tough to predict a traffic problem. The former is only an extrapolation of the railroad. The latter is something completely novel and unexpected."[101] In this narrative, the counterfactual world must be the closest possible to our world, except for the introduction of a technological novelty.

Asimov believed there was some tendency to dislike technological innovations. He illustrated his belief by mentioning Simon Newcombe's mathematical proofs about the impossibility of airplanes. Newcombe was an eminent astronomer. Even after he learned that the Wright Brothers had flown at Kitty Hawk, he adjusted his beliefs only minimally. He stated then that it would not be possible to carry more than one man. Stuart Russell frequently mentions an anecdote analogously to the Newcombe dislike.[102] On September 11, 1933, the father of atomic physics Ernest Rutherford warned, "to those who seek a source of power in the transmutation of atoms—such expectations are the merest moonshine."[103] The very next day, Leo Szilard created the chain reaction.

It is not required for such envisaged circumstances to consider the intermediate steps that bridge the present to the future. In this sense, the imaginary future is not always expressed as a plan but as a prophecy. And if there are no firm grounds that ensure such progression, from the present to the conjectured future, it may only be justified as a value. It is hard to tell if foresight was a prophecy or a plan, or if it was irrational or rational, since the future outcomes are the ultimate judge of this matter, and so these are ex-post-facto statements.

The development of new technologies leads to the conception of dramatic scenarios that follow from "unforeseen consequences" of such developments. Of course, in parallel to those unexpected scenarios, different people try to conceive at least some of them based on logic and sometimes probabilistic reasoning. So, it is precisely this attempt to imagine, account, measure, and predict such undesirable and even potentially catastrophically scenarios that underlies the perception of risk related to them.

CULTURAL REPRESENTATION, AMPLIFICATION THROUGH MEDIA, AND REAL DEVELOPMENT

We propose an adapted framework inspired by SARF for considering the formation of the perception of risk regarding AI, which started with the counterfactual and fictional thinking represented in science-fiction culture. As we will see, the story of risk perception of AI was slowly shaped in the cyclical cultural tripod: *representation—real scientific development—experts' public positioning*.

The fictional thinking has a particular interest in drawing counterfactual scenarios on unknown or obscure things. The power of creation, either derived from God or the human, has always been a topic of curiosity. A possible intelligence or creational ability produced by something or somebody besides God or humans are the centers of gravity of the science-fiction pieces that have represented AI in the past century.

As we will see, with the advancement of the cultural industry from the 1930s and on, there is a proliferation of cultural pieces representing nonhuman intelligences. The dystopian scenarios instigated the public in general, and the interest on it grew *pari passu* with the real development of what further would be called as "artificial intelligence." At some point, in the middle of the path, cultural media and regular (journalistic) media came across on the topic of AI. The same happened with science-fiction writers and scientists (experts) working with AI. The interplay among media, cultural representation, and experts form the circle of the formation of the risk perception of AI.

The interest for the unknown craft counterfactual scenarios is adequately represented culturally in science-fiction pieces, which, in turn, had a massive impact in society, sowing the interest of laypeople and poking the positioning of experts. The media buzzing caught some experts on their topic of research. Some of them get out of the cave to public positioning themselves by also drawing scenarios and shaping perceptions. More than that, they even furnish ideas to science fiction and become literary or movie critics on these topics.

As we will see, both cultural representations and experts often relate AI to risk, something that feeds the interest of the public, governments, and market in an endless circle that intertwines culture, science, and media. In what follows, we present in the next chapters how this adapted framework for understanding the formation and amplification of risk regarding AI has a historical development based on the aforementioned tripod.

NOTES

1. Cambridge Online Dictionary, *Cliffhanger* (Cambridge: Cambridge University Press, 2008). https://dictionary.cambridge.org/pt/dicionario/ingles/cliffhanger
2. Homer, *The Odyssey*, trans. Alexander Pope (London: Bernard Lintot, 1725–6), 12.469–70.

3. Charles Du Cange, *Glossarium ad Scriptores Mediae et Infimae Latinitatis, in quo Latina Vocabula Novatae Significationis, aut Usus Rarioris, Barbara & Exotica Explicantur, Eorum Notiones & Originationes Reteguntur* (Lutetiae Parisiorum: Ludovicum Billaine, 1678).

4. John Florio, *A Vvorlde of Wordes, or Most Copious, and Exact Dictionarie in Italian and English, Collected by Iohn Florio* (London: Arnold Hatfield, 1598).

5. Jacob Grimm and Wilhelm Grimm, *Deutsches Wörterbuch* (Leipzig: Hirzel, 1854).

6. Niklas Luhmann, *Risk: A Sociological Theory* (New York: A. de Gruyter, 1993).

7. Thomas Blount, *The Glossographia: or A Dictionary, Interpreting All Such Hard Words of Whatsoever Language, Now Used in Our Refined English Tongue with Etymologies, Definitions, and Historical Observations on the Same. Also, the Terms of Divinity, Law, Physick, Mathematicks, and Other Arts and Sciences Explicated. Very useful for all Such as Desire to Understand What They Read* (London: Newcombe for George Sawbridge, 1661).

8. For more details, please access the digitalized version of the dictionary: quod.lib.umich.edu/e/eebo/A28464.0001.001/1:7.17?rgn=div2;view=fulltext.

9. R. Cotgrave and C. Hollyband, *A Dictionarie of the French and English Tongues* (London: Adam Islip, 1611).

10. The English Reports, *The English Reports* (London: W. Green, 1900).

11. Oxford Dictionary, *Risk* (Oxford: Oxford University Press, 2019).

12. A notable exception comes from the work of Gary Machlis and Eugene Rosa, "Desired Risk: Broadening the Social Amplification of Risk Framework," *Risk Analysis* 10 (1990), 161–168 and Stephan Lyng "Edgework: A Social Psychological Analysis of Voluntary Risk Taking," *American Journal of Sociology* 95 (1990), 851–886 and *Edgework: The Sociology of Risk-Taking* (New York: Routledge, 2005). These works point out the phenomenon of "desired risk," usually related to people aspiring the experience of thrill. The most common example is extreme sports. Lyng called these voluntarily high-risk activities with explicit threat to one's physical or mental well-being as *edgework*. But we can characterize less obvious examples such as crime, whose sense would be closer to the adventurous undertaking. See, Stephan Lyng, "Disfunctional Risk Taking: Criminal Behavior as Edgework," in *Adolescent Risk Taking*, ed. N. Bell and R. Bell (London: Sage, 1993) and "Crime, Edgework, and Corporeal Transaction," *Theoretical* Criminology 8, no. 3 (2004), 359–375.

13. Adrienne Mayor in *Gods and Robots* (Princeton: Princeton University Press, 2018), highlights the mixed feelings of awe, dread, and hope within ancient Greeks in their myths for artificial life. "The mix of exuberance and anxiety aroused by a blurring of the lines between nature and machines might seem a uniquely modern response to the juggernaut of scientific progress in the age of technology. But the hope—and trepidation—surrounding the idea of artificial life surfaced thousands of years ago in the ancient Greek world. Imaginative myths expressed and struggled with the awe, dread, and hope summoned by the creation of animated statues, attempts to surpass human limits, and the pursuit of immortality. This is a discussion one might say that the ancient Greeks began," 213.

14. Hazard has its roots probably from the Persian word *zar*, which meant *dice*. It was translated into Arabic as *az-zahr,* designating both the dice and then the abstract entity of chance—as a metaphor for playing with the dices. As it happened with *risk*, English borrowed the word from the French *hasard.*

15. Christoph Hohenemser et al., "The Nature of Technological Hazard," in *The Perception of Risk*, ed. Paul Slovic (New York: Earthscan, 2000), 169.

16. Hohenemser et al., "Technological Hazard," 169.

17. William Freudenburg, "Perceived Risk, Real Risk: Social Science and the Art Of Probabilistic Risk Assessment," *Science* 242, no. 4875 (1988), 44–49.

18. Expected utility is a predicted subjective value for one of several options. It is calculated as the sum of the utility of every possible outcome. Each outcome is multiplied by the probability of its occurrence. Utility is a subjective, relative measure of satisfaction, especially in economics and game theory. Utility functions plays an important role in artificial intelligence, one can easily find it within AI's textbooks. See: David Poole and Alan Mackworth, *Artificial Intelligence: Foundations of Computational Agents*, 2nd ed. (Cambridge: Cambridge University Press, 2017), chap. 9; Russell and Norvig, *Artificial Intelligence*, chap. 16.

19. Chauncey Starr, "Social Benefit Versus Technological Risk," *Science* 165 (1969), 1232–1238.

20. Starr, "Social Benefit," 1237.

21. Note that Stanley Kaplan pioneered the quantitative methods for risk analysis. See, John Garrick et al., "Stanley Kaplan: Reflections and Papers Published," *Risk Analysis* Virtual issue (2019), https://onlinelibrary.wiley.com/page/journal/15396924/homepage/virtual_issue__stanley_kaplan__reflections_and_papers_published.htm; Stanley Kaplan, "On The Method of Discrete Probability Distributions in Risk and Reliability Calculations—Application to Seismic Risk Assessment," *Risk Analysis* 1, no. 3 (1981), 189–196; Stanley Kaplan and John Garrick, "On The Quantitative Definition of Risk," *Risk Analysis* 1, no. 1 (1981), 11–27.

22. Stanley Kaplan, "The Words of Risk Analysis," *Risk Analysis* 17, no. 4 (1997), 407.

23. Bernd Rohrmann and Ortwin Renn, *Cross-Cultural Risk Perception: A Survey of Empirical Studies* (Dordrecht: Springer Science+Business Media, 2000), 13.

24. Max Boholm et al., "The Concepts of Risk, Safety, and Security: Applications in Everyday Language," *Risk Analysis* 36, no. 2 (2016), 320–338.

25. Boholm et al., "The Concepts of Risk," 320.

26. See, Sven Ove Hanson, *The Ethics of Risk: Ethical Analysis in an Uncertain World* (Basingstoke: Palgrave MacMillan, 2013); Roger Kasperson et al., "The Social Amplification of Risk: A Conceptual Framework," in *The Perception of Risk*, ed. Paul Slovic (New York: Earthscan, 2000), 232–245.

27. Ulrich Beck, *Risk Society: Towards a New Modernity,* trans. M. Ritter (London: Sage, 1992), 45.

28. Baruch Fischhoff et al., "How Safe Is Safe Enough? A Psychometric Study of Attitudes Toward Technological Risks and Benefits" in *The Perception of Risk*, ed. Paul Slovic (New York: Earthscan, 2000), 80.

29. Beck, *Risk Society,* 22.

30. Two other social approaches to risk worthy to mention are Niklas Luhmann's system theory (Luhmann, *Risk*) and Michel Foucault's governmentality—see. Michel Foucault, "Governmentality. Ideology and Consciousness," in *Power/Knowledge: Collected Interviews and Other Essays 1971–1977* (Brighton: Harvester Press, 1980) and "Governmentality" in *The Foucault Effect: Studies in Governmentality,* ed. G. Burchell, C. Gordon, and P. Miller (London: Harvester, 1991), 87–104. According to system theory, risk and dangers are a possible future loss or damage. The former is attributed to a decision, whereas the latter is assigned to an external cause. In Foucault's approach, risk uses probabilistic techniques to predict outcomes as governmental technology. It may also refer to all attempts to govern in terms of a calculated future. For an excellent summarization of sociological approaches to risk, see Jens Zinn, *Social Theories of Risk and Uncertainty: An Introduction* (Oxford: Blackwell Publishing, 2008).

31. Mary Douglas and Aaron Wildavsky, *Risk and Culture: An Essay on the Selection of Technological and Environmental Dangers* (Berkeley: University of California Press, 1982).

32. Mary Douglas, *Purity and Danger: An Analysis of the Concepts of Pollution and Taboo* (London: Routledge, 1966).

33. Douglas and Wildavsky, *Risk and Culture.*

34. This is an old idea one can already find in the work of the sociologist Robert Merton, "The Unanticipated Consequences of Purposive Social Action" *American Sociological Review* 1, no. 6 (1936), 894–904 and earlier in the classic work of Frank Knight, *Risk, Uncertainty, and Profit* (Boston and New York: Houghton Mifflin Co., 1921). As Merton also notices, "unforeseen consequences should not be identified with consequences which are necessarily undesirable (from the standpoint of the actor) [. . .] undesired effects are not always undesirable effects" Merton, *Unanticipated Consequences,* 895. According to Knight, Risky situations are those where the outcomes are unknown but driven by probability distributions known at the beginning. They differ from uncertain situations, which not only the outcomes but also their probability patterns are unknown.

35. Paul Slovic, "Introduction and Overview" in *The Perception of Risk* (New York: Earthscan, 2000), xxxvi.

36. Slovic, "Introduction," xxxvi.

37. Rohrmann and Renn, *Cross-Cultural Risk Perception.*

38. See: Amós Tversky and Daniel Kahneman, "Judgment under Uncertainty: Heuristics and Biases," *Science* 185 (1974); Daniel Kahneman et al., *Judgment under Uncertainty: Heuristics and Biases* (New York: Cambridge University Press, 1982); Thomas Gilovich et al., *Heuristics and Biases: The Psychology of Intuitive Judgment* (New York: Cambridge University Press, 2002); Daniel Kahneman, *Thinking: Fast and Slow* (New York: Farrar, Straus and Giroux, 2011);

39. Tversky and Kahneman, "Judgement under Uncertainty."

40. Tversky and Kahneman, "Judgement under Uncertainty."

41. Paul Slovic et al., "Decision Processes, Rationality and Adjustment to Natural Hazards" in *The Perception of Risk*, ed. Paul Slovic (New York: Earthscan, 2000), 15.

42. A psychometric study is a theory and technique of psychological measurement.
43. Paul Slovic et al., "Rating the Risks," *Environment* 2, no. 3 (1979).
44. Slovic et al., "Rating the Risks," 106.
45. Slovic et al., "Rating the Risks," 106.
46. National Highway Traffic Safety Administration (NHTSA), *Traffic Safety Facts: Research Note* (U.S. Department of Transportation, September 2012).
47. See the next chapter.
48. An astounding example of the bias of imagination is the risks of harm from grizzly bears' attacks in the United States. From 2010 to 2019, there were only twenty-nine fatal bear attacks in North America, see https://en.wikipedia.org/wiki/List_of_fatal_bear_attacks_in_North_America. Roughly, it makes a 0.08 annuals death per ten million persons. Because of the bears' public image of risk, we imagine it is more likely to die from bear attacks than dog attacks. The annual death per ten million persons in dog attacks is ten times higher than the bears (Forrester, Weiser, & Forrester, 2018).
49. Ali Alhakami and Paul Slovic, "A psychological study of the inverse relationship between perceived risk and perceived benefit," *Risk Analysis* 14, no.6 (1994).
50. Melissa Finucane et al., "The Affect Heuristic in Judgements of Risk and Benefits" in *The Perception of Risk*, ed. Paul Slovic (New York: Earthscan), 413–429.
51. Paul Slovic et al., "The Affect Heuristic," *European Journal of Operational Research* 1777 (2007), 1333–1352.
52. Tanya Anandan, "Robot Safety, Everything But Routine," *Robotic Industries Association* (2015-08-20). https://www.robotics.org/content-detail.cfm/Industrial-Robotics-Industry-Insights/Robot-Safety-Everything-But-Routine/content_id/5653.
53. Eugene Rosa et al., "The Cognitive Architecture of Risk: Pancultural Unity or Cultural Shaping?" in *Cross-Cultural Risk Perception: A Survey of Empirical Studies*, ed. Ortwin Renn and Bernd Rohrmann (Dordrecht: Springer Science+Business Media, 2000).
54. Rohrmann and Renn, *Cross-Cultural Risk Perception,* 15.
55. Peter Bernstein, *Against the Gods: The Remarkable Story of Risk* (New York: John Wiley & Sons, 1996), 2.
56. In a somewhat similar approach regarding the construction of scenarios, Stanley Kaplan argues that there must be a point of departure at which some "Initiating Event" happens. That starts the scenario, and it goes until it ends at an "End State" (Kaplan, 1997, p. 416).
57. Starr, "Social Benefit."
58. Tversky and Kahneman, "Judgement under Uncertainty."
59. See, Kasperson et al., "The Social Amplification" and Nick Pidgeon et al., *The Social Amplification of Risk* (Cambridge: Cambridge University Press, 2003).
60. Ortwin Renn et al., "The Social Amplification of Risk: Theoretical Foundations and Empirical Applications," *Journal of Social Issues* 48, no. 4 (1992), 137–160.
61. Luhmann, *Risk.*
62. Kasperson et al., "The Social Amplification."
63. One can find an application of SARF to current artificial intelligence risk perception in (Neri & Cozman, 2019). However, we present some findings in chapter 5.

64. Paul Slovic, "Informing and Educating the Public about Risk," *Risk Analysis* 6, no.4 (1986), 403–415.

65. Slovic, "Informing and Educating."

66. T. Quandt et al., "Fake News," in *The International Encyclopedia of Journalism Studies,* ed. Tim P. Vos, Folker Hanusch, Annika Sehl, Dimitra Dimitrakopoulou, and Margaretha Geertsema-Sligh (New York: John Wiley & Sons, 2019).

67. Hugo Neri and Fabio Cozman, "The Role of Experts in the Public Perception of Risk of Artificial Intelligence," *AI & Society* 35, no. 3 (2020), 663–673. DOI: 10.1007/s00146-019-00924-9.

68. Slovic, "Informing and Educating."

69. Kasperson et al., "The Social Amplification," 222.

70. Slovic et al., "Rating the Risks," 19.

71. Arthur C. Clarke, *Profiles of the Future: An Inquiry into the Limits of the Possible* (New York: Holt, 1984).

72. Georg Simmel, "The Sociology of Secrecy and of the Secret Societies," *American Journal of Sociology* 11 (1906), 441–498.

73. Ian Hacking, *The Taming of Chance* (Cambridge: Cambridge University Press, 1990), 4–5.

74. Baert and Morgan, "A Performative Framework."

75. Neri and Cozman, "The Role of Experts," 11.

76. First explored in Renn et al., "The Social Amplification."

77. Except by the fatal car crashes with Tesla and Uber's autonomous vehicles. See, Neri and Cozman "The Role of Experts"; Jack Stilgoe, "Machine Learning, Social Learning and the Governance of Self-Driving Cars," *Social Studies of Science* 48, no.1 (2017), 25–56.

78. Suvin is an Yoguslavian compared literature scholar specialized in science fiction and emeritius professor of McGill University.

79. Darko Suvin, *Positions and Presuppositions in Science Fiction* (London: MacMillan Press, 1988), 42.

80. Suvin, *Positions and Presuppositions*, 37.

81. Daniel Kahneman, "Varieties of Counterfactual Thinking" in *What Might Have Been: The Social Psychology of Counterfactual Thinking*, ed. Neal Roese and James Olson (London: Psychology Press, 1995), 379.

82. Kahneman, "Varieties," 378.

83. Kahneman, "Varieties," 379.

84. Kahneman, "Varieties," 380.

85. Nilsson, *The Quest for Artificial Intelligence*, 163.

86. Tversky and Kahneman, "Judgement Under Uncertainty," 11.

87. The study was originally published in 1974 and reprinted in 2000 (Slovic, 2000).

88. Herbert Simon (1916–2001) proposed the idea of bounded rationality (see *Models of Man,* 1957). He was an economist, psychologist, political scientist, and one of the fathers of AI.

89. Slovic et al., "Decision Processes," 25.

90. Paul Slovic et al., "Cognitive Processes and Societal Risk Taking" in *The Perception of Risk*, ed. Paul Slovic (New York: Earthscan, 2000), 44.

91. Daniel Kahneman and Amós Tversky, "The Simulation Heuristic," in *Judgement under Uncertainty: Heuristics and Biases*, ed. Daniel Kahneman, Paul Slovic, and Amós Tversky (Cambridge: Cambridge University Press, 1998). For a more formal computational approach for this procedure, see Pearl's work on causality—Judea Pearl, *Causality: Models, Reasoning, and Inference* (Cambridge: Cambridge University Press, 2009) and *The Book of Why: The New Science of Cause and Effect* (New York: Basic Books, 2018).

92. Isaac Asimov, "Social Science Fiction," in *Modern Science Fiction: Its Meaning and Its Future,* ed. Reginald Bretnor (New York: Coward-McCann, 1953), 158–159.

93. Bertrand Russell, *The Scientific Outlook* (London: Routledge, 2001), 193.

94. David Lewis, "Truth in Fiction," *American Philosophical Quarterly* 15, no. 1 (1978), 37–46.

95. Lewis, "Truth in Fiction," 40.

96. Asimov, "Social Science Fiction."

97. Asimov, "Social Science Fiction," 160.

98. Mayor, *Gods and Robots.*

99. Asimov, "Social Science Fiction," 166.

100. Spike Jonze, "Spike Jonze on letting Her rip and Being John Malkovich," interviewed by Chris Michael, *The Guardian*, September 9, 2013. https://www.theguardian.com/film/filmblog/2013/sep/09/spike-jonze-her-scarlett-johansson. Accessed October 4, 2019.

101. Asimov, "Social Science Fiction," 172.

102. See, Russell, *Human Compatible*; Stuart Russell, "Ban Lethal Autonomous Weapons," *The Boston Globe*, October 9, 2015, and "Take a Stand on AI Weapons," *Nature* 521 (2015), 415–416.

103. Richard Rhodes, *The Making of the Atomic Bomb* (New York: Simon & Schuster, 1986), 27.

Chapter 2

We Wanted to Try

Imagined Robots and Mathematical Experiments Pre-1960s (in Collaboration with Veridiana Domingos Cordeiro)

A clear representation of intelligent machines presenting side effects comes to the surface in the 1920s, in particular in European countries in post–World War I (1914–1918). The representation of artificial intelligence started in association with "automata" or in today's words "robots." Automata have been wandering around Europe for centuries and boomed during the late eighteenth century; for instance, Michel Foucault (1926–1984) mentions this fact in *Discipline and Punishment.* In E.T.A. Hoffman's *Der Sandman* (1816), the protagonist fell in love with an *automaton*, Olimpia, and went mad when he figured out that she was not human. The golden age of automata in Europe ranged from 1860 to 1910, when there were families of automata makers especially in France (e.g., *the Bontems*, the *Roullet & Decamps*). This automata culture was not only confined to Europe. During the Edo period (1603–1867), Japan adopted automata known as *Karakuri ningyōm*, as a form of entertainment. However, automata making was not just an oddity at the fringes of the economy there. The founder of the Toshiba Corporation, Hisashige Tanaka (1799–1881), also known as the "Thomas Edison of Japan," was a craftsman who created complex automata—some examples of his creations were tea-serving and *kanji* painters automata.[1]

The cultural representation of intelligent machines took off in the 1920s, even though there were some literary representations of automata before. We underscore three remarkable examples. Samuel Butler's *Erehwon: or over the range* (1872), a social fiction placed in Erehwon (reversed for nowhere), where Erehwonians had the perception that machines were potentially dangerous since they could surpass humans.[2] Gustave Le Rouge's *Metal Man* in *La Conspiration des Milliardaires* (1900) and Frank Baum's copper robot *Tik-Tok* who appeared in the third *Oz books* (1907). *Tik-Tok* was one of the earliest fictional appearances showing true machine intelligence. However, it was not a free agent.[3]

The invention of the cinema and its popularization played an important role in propagating a certain representation of artificial intelligence, which was certainly not cheerful. Early TV shows also depicted robots always within science fictions and they were represented as villains most of the time. At the end of this period, the rise of television popularized some cultural icons such as *Robby the Robot.* Communication via telephone instilled in Americans and Europeans the idea of instant communication. However, before World War II, science-fiction pulp magazines popularized the first representations of artificial intelligence, especially in the shape of robots. The most striking example was Isaac Asimov's robotic laws, which worked as *risk mitigators* and towered the world of explicit fiction to general and sometimes scientific discussions about the oughts of developing artificial intelligence. Furthermore, Asimov applied the laws later to complex informational systems, which indicated how the abstraction of artificial intelligence was brewing. The rise of science fiction sets the representations of what to expect from machines that display intelligent behavior, ranging from positive to negative scenarios—even though the negative scenarios predominated. But it has also cultivated a generation of fans, some of them would work in something related to science. There were fan conventions, and the first Worldcon or World SF Convention took place in New York in 1939; obviously, it was not as a big event as today's events.

The computer was a much later invention, and consequently, its cultural representation was not an object of the domain of writers and philosophers, but mathematicians and engineers. Before the 1920s, we may find a rough depiction of an *ur-computer* in E. M. Foster's *The Machine Stops* (1909), a story where humans live below the ground in isolation and their needs are met by an omnipotent global machine. Also, their communication is made via some sort of instant messaging. The next representations of a computer appear at least three decades later with Robert A. Heinlein's *Misfit* (1939), A. E. van Vogt's *The World of Null-A* (1945), and Asimov's *Escape!* (1945). The scene changed completely during the 1950s with the advent of computers made from transistors instead of vacuum tubes. It certainly changed how one could represent a computer. During the vacuum tubes' era, the computer was more like a spatial entity, occupying an entire room. The quick progress in the computer hardware led to the miniaturization of computers. This swift shift was probably felt more strongly than we can imagine today. It may have happened for those who dealt with computers in their routines. We still have an analog situation today in which data centers are these sorts of rooms filled with closet-like computers; however, just a few people had any contact with this sort of equipment.

With computers and computer systems, one has more elements to conjecture future consequences. Alan Turing had such a vision as well as Norbert Wiener. Nonetheless, both were mathematicians. They were examples of experts involved in artificial intelligence who could make a public representation of future risks of it. Wiener envisioned the risks from artificial intelligence and made public interventions about it. In our model of risk, he was a "risk communicator." In some sense, he framed a more plausible scenario about how things can go wrong, attenuating more unrealistic and dystopian representations of risk from that time. Finally, it is important to highlight in the risk front that nuclear weapons dominated actual and imaginary concerns, especially in the Cold War context. Thereby, the association and attempt to apply any technology in the military domain were always on the horizon, and the funded research bend toward this direction.

Finally, already at the end of this period, during the 1950s, the representations, ideas, and technological developments regarding kinds of nonhuman intelligence are named for the first time. In 1956, in the "mythical" Dartmouth congress, John McCarthy coined the expression "artificial intelligence." Although it may look just an anecdote, it created unity and identity to the variety of researchers going on, as well as virtual scenarios where machines present some degree of intelligence. Moreover, the term enabled the thinking that human-made intelligence could be not only embodied in a robot but disembodied in an abstract system. In this sense, the term "artificial intelligence" represented an epistemological stepping stone.

REPRESENTATIONS: FIRST ROBOTS, *FROM SLAVES TO DEVILS*

Rossum Universal Robots—R.U.R (1921)—is a book written by the Czech Karel Capek. He coined the term "robot." In Czech, *robota* means "forced labor," which is related to the word *rab*, that is "slave." The term emerged to name machines produced in a factory of "artificial people," made from organic matter instead of metal. This means that they were composed of flesh and blood, and they had their own intelligence. However, they were slaves driven for and by regular humans. Although they looked happy in their activities, they gathered together in a rebellion against ordinary humans. In the very first moment, the term "robot" was closely connected to the idea of risk, fear, and unforeseen consequences. Asimov calls this typical representation of artificial intelligence as "the Frankenstein's paradox". Later, Capek's book was adapted into a theatre play and translated to thirty languages. Before the word "slave" is directly related to the concept of "robots," Frank Baum's

Tik-Tok made this relationship explicit some years before, with robots as servants with no emotions.

In the same year, the first Italian fictional movie, André Deed's *L'uomo mecanico* ("The mechanical man") pictured a story of a robot projected by a scientist that walks freely around the streets but is remotely controlled. With its strength, the robot was capable of destroying houses and scaring people. During the movie, the robot is introduced in a social party, where he and the people around seem to be happy. However, as soon as a gang kills the inventor, the robot swiftly gets out of control. The malefactors drove the robot to commit a variety of crimes, including murdering. At the end, the inventor builds another model to fight the criminal robot. This first cinematographic piece has already represented a nonhuman intelligence as a malicious tool, a robot servant in an anthropomorphic shape that cannot distinguish lawful from unlawful actions. Unfortunately, the full version of the movie was lost; there is a unique shortened version at *Cinemateca Brasileira* ("Brazilian Cinematheque").

Still, in the 1920s, the most relevant sci-fi movie with artificial intelligence's representation in-between war's period was Thea von Harbou and Fritz Lang's *Metropolis* (1927). According to the critics Workman and Howarth (2016), Fritz Lang or Thea von Harbou likely have been influenced by Deed's *The Mechanical Man.*

The American audience acknowledges the movie's technical marvel, as the *New York Times*' review from March 1927 states:

> It, therefore, stands alone, in some respects, as a remarkable achievement. It is a technical marvel with feet of clay, a picture as soulless as the manufactured woman of its story. Its scenes bristle with cinematic imagination, with hordes of men and women and astounding stage settings. It is hardly a film to be judged by its narrative, for despite the fantastic nature of the story, it is, on the whole, unconvincing, lacking in suspense and at times extravagantly theatric. It suggests a combination of a preachment on capital and labor in a city of the future, an R. U. R. idea and something of Mrs. Shelley's "Frankenstein." Its moral is that the brains and the hands fail when the heart (love) does not work with them. The brains represent capital, and the hands, labor. The production itself appears to have been a Frankenstein model to the story.[4]

In 1947, the German sociologist Siegfried Kracauer summarized the national reactions toward the movie, "Metropolis impressed the German public. The Americans relished its technical excellence; the English remained aloof; the French were stirred by a film which seemed to them a blend of Wagner and Krupp and, on the whole, an alarming sign of Germany's vitality."[5] Wagner was the famous German composer, and

Krupp was an armament manufacturer. Kracauer also added the critics of the champion of science fiction, H.G. Wells, who thought the movie as "quite the silliest film," and Conan Doyle, who was unenthusiastic about the movie.

Besides the critics, *Metropolis* was good social-science fiction. As Kracauer evaluated in the context of German cinematographic history,

> In their concern with the aims of rebellion [. . .] One film was more explicit "than all others: METROPOLIS. In it, the paralyzed collective mind seemed to be talking with unusual clarity in its sleep. This is more than a metaphor: owing to a fortunate combination of receptivity and confusion, Lang's scriptwriter, Thea von Harbou, was not only sensitive to all undercurrents of the time, but indiscriminately passed on whatever happened to haunt her imagination.[6]

The *Maschinenmensch*, the robot in Metropolis, played a role already presented in German movies. A key example was the Wegener's *Der Golem/The Golem* (1915), a fantasy movie in which the first proto-robot appeared on the screens. The story stems from the medieval Jewish tale of the Rabbi Loew of Prague, who infused life into a clay creature. The creature turns against the master, who was an antique dealer. In *Hommunculus* (1916), a famous movie during the Great War, the artificial intelligent *automaton* becomes conscious and plots against humankind. He did it through deception. "Disguised as a worker, he incites riots which give him, the dictator, an opportunity to crush the masses ruthlessly. Finally, he precipitates a world war."[7] However, the automata's side effects in *The Golem* and *The Mechanical Man* differ from the ones in *Homunculus* and *Metropolis*. In both *The Golem* and *The Mechanical Man*, the robots turn against humans in a "sentimental rush," taking advantage of their higher physical power. In the *Homunculus* and *Metropolis*, the robots' plots anticipate humans and cause social distress. At least in *Metropoli*s, the robot is told to do so. In this sense, it resembles the perfect tool for malevolent acts. When in action, this malign representation of artificial intelligence is a technological representation of the devil's figure. They are diabolic in the strict sense of the Greek word *slanderer*.

In the next decade, the early 1930s, the "Professor Jameson" series by Neil R. Jones featured human and alien minds preserved in robot bodies. It was further republished in the 1960s, as showed below. Although the cover is from the 1960s, it represents very well the idea of the book. It still did not converge body and robot in a single entity. The story keeps the representation of something purely human that can embody into a machine.

Even Superman emerged side by side with robots in 1941, when it was released the animated short *The Mechanical Monsters*, in which the hero features battling against a mad scientist, who "owns" a small army of robots at his command. Again, the breathing-fire robots are pictured as mass destruction machines who rob banks.

A more complex representation of artificial intelligence, its impact in society, and consequently its risks, came to the public via Isaac Asimov. He was a prolific writer. His career started writing short stories in 1939, and he has never stopped. However, his work in the so-called "golden age of science fiction," during the 1940s and the early 1950s, that sustained his reputation later. Of the utmost importance, Asimov created the pivotal representation of future risks from robots until the present days. He developed this representation in his *I, Robot* (1950) stories. Before the 1950s, he was one of the first writers to present a computer-based AI character in *Escape!* (1945), represented in the story as "*the brain*", a supercomputer with an infantile and human-like personality.

Like many other science-fiction stories during Asimov's period, his collection of nine stories were published in science-fiction magazines, the *Super Science Stories*, and *Astounding Science Fiction*—we will see more about this media in the next section. The set of stories creates a narrative arch from the beginnings of robots, when they couldn't speak, to the end when they stand between humankind and its destruction.

As Darko Suvin evaluates, the robot series is a

> "future history" [. . .] based on two premises: first, that except for one factor, human behavior and the social system—for example, press reporters and giant corporations—will remain unchanged; second, that the new, change-bearing factor will be the epoch-making technological discovery of "positronic brainpaths," permitting mass fabrication of robots with intelligence comparable to human.[8]

The Robot series is social-science fiction of the very first rank. Asimov did it conceiving the closest possible world after the introduction of a cognitive or technological *novum:* the mass-produced artificial intelligent robots. In the sequence, he examined its impacts on human behavior and institutions. His celebrated "Three Laws of Robotics," responsible for constraining the robots' actions, was first made explicit in *Runaround* (1942). It was formulated in collaboration with John W. Campbell, who was also editor of the famous magazine *The Astounding Science Fiction.* Campbell also believed in the inevitability of human superseding by intelligent machinery. The laws are the following,

1. A robot may not harm a human being, or, through inaction, allow a human being to come to harm.
2. A robot must obey the orders given to it by human beings, except where such orders conflict with the First Law.

3. A robot must protect its existence as long as such protection does not conflict with the First or Second Law.

Although the three laws are the golden nugget of his work, the robot series explores a more subtle consequence of developing artificial intelligence, such as the human antipathy against machinery and the conflicts that arise from there. For example, the first robot novel *The Caves of Steel* (1954) is a sci-fi detective story in which an eminent scientist has been murdered on Earth. The scientist was a Spacer, the fictional affluent first human to emigrate to space. Earth detective Lije Bailey is tasked to investigate the crime, assisted by a Spacer robot called Daneel Olivaw. Unlike the Earth robots, which are metallic and so artificial, Olivaw is constructed to resemble precisely a human being. Bailey was suspicious of the Daneel and is initially hostile to it. He even suspected and accused the robot of the crime, which turned out to be absurd; since no robot could murder a human being because of the three laws.

As a matter of fact, the three laws worked as a risk mitigator factor for the development of artificial intelligence. Adam Roberts argues that Asimov imagined artificially intelligent robots as not only humans but in many ways as "more human than humanity,"[9] unlike the previous representations of clear danger, embodiments of the threat of technology. The laws worked as a built-in ethical system that they were bounded to obey. Therefore, the robots were more moral beings than humans.

Asimov devised an ethical programming in which robots have Kant's moral imperatives hardcoded. This is an absolute instantiation of Kant's idea, "*the starry heavens above me, and the moral law within me.*"[10] With his robots, Asimov created a race of sentient, thoughtful beings in whom the Kantian moral imperative is internalized. Robots did not consult their conscience when they faced an ethical dilemma. They obey the three laws that fully drive their behavior. Almost all the robot's stories contain the working-out of the implications for what it would be like to live life under such imperatives. The stories also explore the consequences of moral constraining. In *The Evitable Conflict* (1950), the machines in control of the Earth's economy figured out that the only way to comply with faultlessly with the First Law was by taking control of humans.

With some suspicious, Darko Suvin and the writer Stanislaw Lem say that "his ingenious mimicry of the Decalogue and the Kantian categorical imperative in the form of Newtonian laws cannot, therefore, be taken at all seriously as a basis of prophetic extrapolation."[11] But they can be imagined, and as we have learned with David Lewis before, if it is true in Asimov's world, that is enough. It is worthier if we think counterfactually on Asimov's scenario. If the laws were erased, it would take the lid off the potential threat that artificially intelligent robots can bring. That is precisely the cliffhanger idea of the laws that arise in the robot stories. Furthermore, that is the reason why some humans are suspicious of them. It creates the tension that any human-made law will eventually be broke or wrest it.

In the 1950s, the sociopolitical context shapes the background of these narratives. Space comes into the game and the robots interplay with this new scenario. Herman Cohen's movie *Target Earth* (1954) still does not encompass the space itself; however, it shows spatial robots coming from Venus and invading Chicago, which was swiftly evacuated. The death rays of the robots increasingly kill the few survivors. The use of atomic weapons is already contemplated in this movie—curiously in the same city of the world's first artificial nuclear reactor, which was demonstrated at the Metallurgical Laboratory at the University of Chicago in 1942.

In the same way, on the other corner of the world, in Japan, cartoons, specifically "*animés,*" were also successful in representing robots and blurring the line between man and machine. In 1951, Osamu Tezuka, a renowned *animé* artist, created the "Tetsuwan Atomu" (later translated as *Astro Boy*), a fictional series in which a futuristic environment is a scenario where the story of the robot Atom plays out. The mighty robot Atom, created by the Ministry of Science and scientist to replace his dead son, is treated as a family member. After realizing that Atom could not appreciate flowers, but only the mechanical forms of a cube, the Ministry of Science decided to sell him to a circus. When the circus was on fire, the robots, including Atom, saved the people. The Ministry of Science saw a core human's trait in this attitude and decided to grant Human Rights for all the robots in Japan. After a while, Atom starts to express emotions and become a leader in the fight against injustice and crimes. Although Atom used to fight against villain robots, humans held him in high esteem. Some years later, Atom got his own robotic family. In the 1960s, Mitsuteru Yokoyama created the *mangá Tetsujin-28*, the story of a ten-feet tall super robot. Under the command of ten-year-old boy Shotaro, the son of the scientist who created the *Tetsujin-28*, the robot fights against crimes, aliens, and evil robots.

In these early representations of artificial intelligence, the plot follows a narrative that begins with a scientist capable of creating a machine that serves humans in their needs. Still, it likely gets out of control, becoming a threat to society. In the Western representations, the machine power is seen as something potentially catastrophic, which may come out of control. Consequently, artificial intelligence is rarely depicted as "one of us" at odds over Japanese representations. Such as in Europe and the United States, robots' power of destruction is also acknowledged; however, many robots are tamed. Their power is used to fight against other robots (when necessary) and help humans. They are helpmates and faithful companions for children since the beginning. We will try to explore this cultural variation in the next chapters.

Why there is a demonization's bias of machines in science-fiction stories in Western countries?

> In philosophical terms, the machines are seen as inherently less authentic than organic life because they fall under the rubric of techné rather than episteme. Good means amenable to humanization, like Asimov's saintly Bicentennial Man; bad means resistant to this process. More recent SF has been bolder in deconstructing this notion, with a range of cyberpunk and other texts exploring the validities of a technological perspective, but the bulk of the genre reproduces the ancient bias.[12]

It is worth to note that, since the beginning, the representation of a possible nonhuman intelligence has been directly linked to an anthropomorphic machine, a robot. The threatening aspect of the robot surpasses most of the Western literature and cinematographic pieces. The risk representation of machines getting out of control is just around the corner in most of the narratives. Thus, from slaves, robots may become killers. At this first chunk of time outlined here, at least in the West, the robots still were not represented as holding spirit, emotions, ethics, or any other human trait. The imagination again bumps with the most profound social and religious aspects. For handling with that, the cultural representations either conceive them as evil beings who could destroy the world, either assuming a complete dissociation between mind and body (machine, actually).

MEDIA: THE BIRTH OF CULTURAL INDUSTRY

In these first decades, there was a variety of cultural products from many different countries. As we will see, after the World War II, with the raising of the United States as a sturdy leading country and its leading cinematographic industry, most of the representations of the Western world followed the American books and movies. The growth of the American Industry converges with the emergence of the cultural industry itself, which, in its turn, coincides with the raising of science fiction as a celebrated genre worldwide.

As John Rieder argues, science fiction has been occupying a central place in the cultural consciousness since the late nineteenth century.[13] It raised to the popular imaginary *pari passu* with the emergence of mass culture. The mass culture is the popularization of culture and the dissemination of a variety of media to make it more accessible and comprehensible to laypeople. The *vaudeville* came to amplify and massify the theatre, the *operetta* emerged as a pocket form of the regular operas, and the posters became a reproducible form of paintings. All of these products were captured, reproduced, and conveyed by the cultural industry that emerged to provide entertainment for the audience, which now is converted into "consumers".[14]

The same happened with the readership. The creation of a mass literary culture was tied to some social and cultural changes that trace back to the end of the nineteenth century. As books were in general costly items, few titles came out each year. Magazines and paperback titles tried to fulfill this blank. These media served as platforms for a considerable amount of stories to be published. The emergence of a specialized market for science fiction served as a cushion for elaborated cultural representations on the risk of technology in general. Before 1926, no significant group of science-fiction writers as professionals entirely dedicated to producing stories could have existed. The pulp magazines were the main factor for changing the industry; in particular, Hugo Gernsback's *The Amazing Stories* was a watershed for the specialization of the market. The most prestigious prize for a science-fiction work, the Hugo prize, comes after Gernsback's name. The magazines were the entrance for science-fiction writers. With the relative success of the *Amazing Stories*, different science-fiction magazines were out. The market for science-fiction books was reduced or nonexistent until the 1950s.

The stories broadcasted on radio also played an important role. Radio was the most important cultural propagator of any representation of risk of that time. The most emblematic event was the Orson's broadcast of an adaptation of H.G. Wells's *The War of the Worlds* in 1938. It is told that this broadcast caused national hysteria—however, such statement has been challenged recently.[15] The cinema had also an important role, especially in the United States, the place for science fiction as an autonomous market.

THE MAGAZINES

The magazines were the first medium to convey science-fiction stories. It started with the pulp magazines in the United States, the United Kingdom, Germany, Spain, France, and the Union of Soviet Socialist Republics (URSS). However, it was largely an American phenomenon. Pulp magazines, popularly known as "the pulps," were low-price fiction magazines published from 1896 to the late 1950s. The term "pulp" referred to the quality of these magazines' papers, which allowed their costs of production to be low.[16] The pulps also brought the expression "pulp fiction" into pop culture, a reference to its run-of-the-mill, low-quality literature. The plots of these stories were known for their lurid, exploitative, and sensational subject matter, even though many respected writers wrote for pulps. Adam Roberts, a science-fiction writer and scholar, says that "all the pulps traded in more or less explicitly coded sexual fantasy."[17] The covers, drawings, and pictures published in the issues were as important as the stories. Sometimes the covers would be even designed first. It was the hallmark of science-fiction imagery. Regardless of the variety of covers, the typical pulp magazine cover is represented in common sense as a depiction of a heroic man, attractive women in a state of undress and monstrous aliens, robots posed in mid-action or detail monumental futuristic technology. Robots were depicted as villains in oft dualist narratives.

Before World War I, the best-selling magazine titles *Argosy*, *Adventure*, *Blue Book*, and *Short Stories*, plus the other four, had the combined monthly circulation of more than 2.7 million copies—an estimated readership of 15 percent of the U.S. population. The popular readership market was emerging as a profitable business enterprise. The economic success brought imitation and specialization of the magazines, even though the plots of the stories looked similar with few exceptions. By the 1920s, there were pulp magazines for crime, Western, adventure, romantic also general magazines. At their peak of popularity, the pulp magazines could sell up to one million copies per issue. Thus, science-fiction stories jumped into this market.

The commercialization of science-fiction stories meant that there was some public ready to consume science or scientific and technological related topics *as* entertainment, a trend well captured by the science-fiction magazines, by Hugo Gernsback in particular. This movement started about a century earlier when they derived from the first science magazines aimed at a public of enthusiasts. Magazines were a place where a scientist or a philosopher could easily conceive scenarios based on his or her present knowledge of current technology. In the UK, the mother of all these science and (latter) science-fiction magazines, *The Monthly Magazine* (1796), for instance, had contributions from Thomas Malthus.[18] He was one of the first intellectuals who thought seriously about future trends. Also, he projected that the population growth would outstrip agricultural production according to his knowledge at the moment.

The real big step for magazines was George Newnes's *The Strand Magazine* (1891) in the UK, the first magazine of its kind priced just six-pence—Newnes was one of the founding fathers of the modern journalism. *The Strand Magazine* sought for easy-to-read short fiction stories, a variety of general interest articles and many black and white illustrations. Before *The Strand,* most British magazines printed solid pages of text with few artworks in the middle. It published stories from now high-esteemed writers such as Arthur Conan Doyle's *Sherlock Holmes* and *The Lost World* and H.G. Wells's *First Men in the Moon*. Other magazines that came by following the success of *The Strand* also published serialized classic science-fiction stories. For example, *The Pearson* published Wells's War of the Worlds (1897). Moreover, much of the significant work of Jules Verne appeared in those magazines.

In inland Europe, Germany led the market in science-fiction pulp through the 1920s. As we will see, Germany had a lively science-fiction thrust in this period, and it could represent a broad audience the dread from the development of a robot with artificial intelligence. During the Nazis' rule, such imaginary lavishness was not encouraged. In France, the nineteenth-century popularity of science fiction continued until World War II. In Fascist Spain, there was a boom of science-fiction pulps led by "Los Vampiros del Aire," by José Canellas Casals and Francisco Darnis, a story of a private group of crime fighters who use flight technology to come after criminals. Finally, in post-Revolutionary Russia, "there was an explosion of popular science fiction, seen by the authorities until the later 1930s as a legitimate vehicle for stories expressing the superiority of communist ideals over capitalist ones."[19] In the URSS, these magazines enjoyed enormous readerships.

In the United States, Gernsback, who already had other magazines related to science and technology, launched the first science-fiction pulp magazine in 1926. Other sister and competitors magazine followed Gernsback, *Science Wonder Stories*, *Astounding Stories of Super-Science*, *Wonder Stories*, *Astounding Stories*, *Thrilling Wonder Stories*, and *Flash Gordon Strange Adventure Magazine*. These science-fiction magazines not only survived, but they also proliferated during the Great Depression.

During the first years of the magazines, reprints and amateur writers dominated the scene since there were no science-fiction writers. The reprints were mostly "classical" literature by H. G. Wells, Jules Verne, and Edgar Allan Poe. Regarding the amateur writers, they started as part-time specialists. During the first years, six men produced most of the stories. Professional writers appeared after 1931. The American editor and science-fiction scholar, Everett Bleiler, provides some (not so accurate) data about the occupational information of the science-fiction pulp writers from 1926 to 1936.[20] It is known that 510 writers collaborated on the magazines with science-fiction stories during these periods. The data relate to 222 writers. Almost 40 percent of them were writers or editors and writers, around 15 percent were journalists, 11 percent engineers, 8 percent businessmen, and only 5 percent scientists. The vast majority of authors were American and few Canadians.

Even though we know that the audience was wide, there is no accurate information about the science-fiction magazines' readership. The estimative is that over the years from 1926 to 1936. There were 25,000 to 60,000 or more monthly readers.[21] The audience changed from more mature to younger readers. The subset of mature-reader probably was acquainted with Gernsback's technical magazines (on the radio, electronics, and inventions). As Bleiler informs from his research, the members of fan clubs "were overwhelmingly male and almost totally White, with only a single Native

American in California. No Blacks or Asians seem to be known."[22] The price of the magazines lowered the barriers to the consumption of the stories.

Mature audiences sought social fiction, which was "essentially an extremely mature fiction."[23] As we recall, social fiction was the parent node of the younger science fiction. During the magazines' period, adolescents entrenched the front of the science-fiction audience. The magazines' editors and writers had to accommodate their narratives for the young audience's taste and understanding. Otherwise, they would not be profitable. Thereby, the Western-style adventure thrived. We cannot just oversimplify this period. Also, we cannot ignore exceptions. After all, dealing with risk is not dealing with the norm. As we saw in the Introduction, *The Thought Machine* story came out at the beginning of the pulp magazines. Remember the strong Nietzschean influences in the narrative. It does not seem that the anonymous writer designed the story having in mind a too young audience.

The composition of the science-fiction audience those days may explain why the plot of the stories did not contain major political-economic events of that time. From 1926 to 1936, the magazines were exposed to the Great Depression and the rise of Fascism. However,

> there are very few references to joblessness or economic misery, and even these few are muted. No heed is paid to the social disorders, like food riots or the attacks on the bonus marchers, that accompanied the Depression, to the slogans of the Hoover administration, or to the later government programs like the New Deal, the CCC or NRA. "In Caverns Below" by Stanton A. Coblentz . . . "There are no references to the Spanish Civil War (beginning 1936), the Italian invasion of Ethiopia (beginning 1935), the civil war in China (except for John Taine's 'White Lily'), civil disobedience in India, the German abandonment of the League of Nations, the Japanese invasions of China, the Young Plan, the Nuremberg Laws, or the other really important international events of the time."[24]

According to Bleiler, it might happen because of the editorial tabu, and the pure entertainment was over-appreciated, "in this sense, science-fiction was a cut-off from the realities of the day, but not quite escapism."[25] At least, the social problem gangsters and the mafia were frequently portrayed. Moreover, many concepts were not entirely derived from science but "created on a mythical level."[26] We may speculate that some of the early generation artificial intelligence researchers might have been exposed to such culture. Marvin Minsky, for example, mentions the impact of science fiction in his childhood. "A lot of things I did came from [. . .] as a child, reading science fiction [. . .] when I was a child I ran across H.G. Wells [. . .] I certainly read the English translation of Jules Verne and the short stories of H.G. Wells which had all sorts of hyper-modern ideas in them."[27] Since he was born in 1927, he probably knew these writers via the pulp magazine translations. He also got in contact with Isaac Asimov's and Robert Heinlein's stories, which were heavily published into those venues,

> Just when I'd read all the classics of which there are only about 10 or 20, then another [. . .] an Isaac Asimov appears or [. . .] and a Robert Heinlein—1940—and that renewed my interest and these people also had short stories and longer ones with new ideas about possible new sciences and new ways of looking at things and [. . .] the end result of that is that generally I'd read very little except science fiction and science.[28]

As the young Minsky at that period, many other experts came from that pool of science and science fiction. What is known about a tiny population of science-fiction active fans, some important people appeared from there, such as Isaac Asimov, Arthur C. Clarke, John Robinson Pierce (who was a director of the renowned research lab *Bell Labs*—the "Google Brain" of old times), Frank Kelly (speechwriter of President Truman and founder of the Nuclear Age Peace Foundation), and Martin Gardner (long associated with the *Scientific American* and the popularization of science). These kinds of people were exposed to the declared enthusiasm from editors such as Gernsback and John W. Campbell (1910–1971), the editor of *Astounding Stories*, who were the public defenders of science at the period. The former, in particular, defendedthe importance of the role of science fiction not only for science but also for society. He did it through the editorials of the *Amazing Stories*. Bleiler summarizes it well,

> It made strange claims. It claimed paternity from science, whence its name, and it promised (according to its more enthusiastic promulgators) to influence and possibly to create the future. . . . [He] advanced a theory whereby science-fiction displaces and offers an enlarging and emending critique of science and technology, creating a new synthesis. . . . This is, of course, farfetched and somewhat incredible, but it embodies the messianic thought accompanying early genre science-fiction. Despite its pulp clothing, Gernsback's science-fiction shared a little with evangelical literature and social protest, though it was neither religious in a traditional sense nor utopian. It thus has points of interest beyond a gathering of implausible stories.[29]

In spite of Gernsback's communication of technology appeared as a progressive instrument for humanity, beneath this surface, the stories portrayed in the science-fiction pulp magazines were "really a literature of catastrophism and despair."[30] Themes of world destruction and human devastation were cardinal, and the apocalypse was always on the horizon.

Bleiler categorizes in six story modes all the pulp science-fiction stories published in the United States from 1926 to 1936: (a) conflict (actual physical violence and personal confrontations); (b) crime and detection (crossovers from contemporary detective story, and some space pirates stories), (c) eros (sexual material that goes beyond incidental involvement with women), exploration (general stories of adventure), (e) exposition (description of why and how something works, utopias fit in this category), and (f) things-go-wrong (side effects and conclusive tragedies).

"Things go wrong" is an ultimate, far more than in other types of pulp fiction, with a fair amount of them. Around 30 percent of all stories concerned about future risk of processes, inventions, actions, or social matters that bear their own inner destruction or malfunction. The largest groupings of stories are "things-go-wrong" and "conflict." Bleiler also tabulated the central motifs of the stories. Among the thirty-two motifs of his classification, we are interested in two of them: "cyborgs" and "machine intelligence." In our actual terms, "robots" and "artificial intelligence." They represented around 4 percent of all stories during this period. *The Thought Machine* we presented in the Introduction is an example of such stories. Aliens' related motifs, along with space travel and space operas, represented more than half of the stories. During some years, they could compose about three-fourth of the overall stories. Other distinct themes were medical materials and inventions. As we can see, up to the late 1930s, robots and artificial intelligence were not so popular motifs. We should not forget that the computer was not around yet. Despite that, these stories inherited from the general science-fiction narrative, the dramatic effect of tragedies, calamities, and doom.

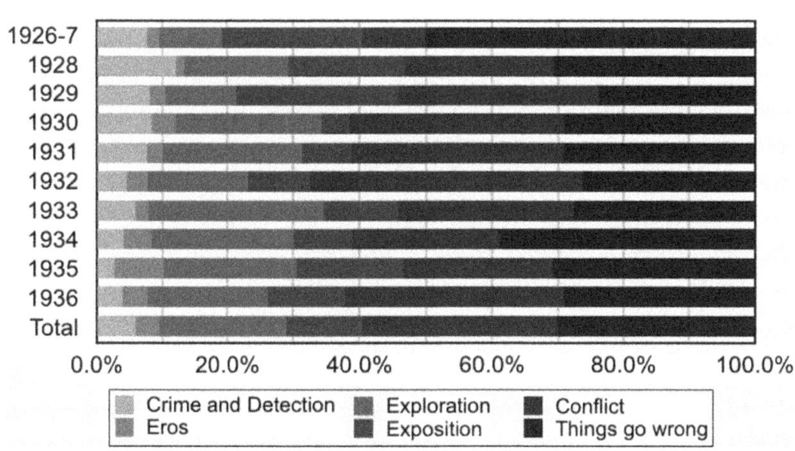

Figure 2.1 Categories of the Science Fiction Pulp Stories. Format by the author. *Source* Bleiler, *Science Fiction*.

In sum, science-fiction narratives showed the tendency to depict the potential undesired scenarios we may fall into either because of a particular instrument or by the consequences these instruments may bring. In this vein, it resonates with the cyberpunk movement. Even though the stories were portrayed for cheap consumption, the act of exploring the consequences of technological development in the future may lead us to overestimate calamitous scenarios. Among many other things, such as the oddity of some science-fiction elements portrayed in the stories, indeed, the uneasiness feeling was arisen by science-fiction stories on the public. The 1920s and 1930s were periods of social realism in general thought and literature. The general American public stigmatized science fiction, deeming it "crazy stuff," something impossible, frowned upon, and undesirable as an anecdote illustration of this situation that can be found in the autobiography of the science-fiction writer Jack Williamson. When he was undergoing thorough analysis at a Clinic in the 1930s, a doctor commented that writing science fiction was a symptom of neurosis. Differently, today, a few cannot empathize with a science-fiction movie or television show.

The pulp magazines declined in the 1940s mostly because of the heavy competition from other media such as the comic books, the television, and the paperback novel conveyed within the cultural industry. Moreover, in a more affluent post-war America, the price gap to more expansive magazines lowered. The price was a critical factor in society just recovering from the Great Depression. "The essence of pulp Science Fiction is not its magazine format, but its cheapness."[31] Pulp magazines epitomized the ephemerality of popular twentieth-century print culture. Their success also correlated with mass production and marketing techniques. It also left its offspring up to these days, especially with the superhero's culture. Modern comic books are sometimes considered descendants of "hero pulps" magazines, which often featured illustrated novel-length stories of heroic characters, such as Flash Gordon.

Some science-fiction movies were shot during the silent era. We have already mentioned the Italian film *The mechanical man* (1922) and the German film *Metropolis* (1927). Sound changed the way cinema worked, but sci-fi would be willy-nilly anglophone biased since the most significant movies were produced in the United States or the United Kingdom. The devastation of Europe by World War II lagged behind former countries producers of science fiction, especially Germany, and enabled the dominance of the American style of movie production in its wake.

The shallow Western-like sci-fi dominated. Even though there were some reasonable attempts to break with this tradition before the 1960s (e.g., George Pal's *Destination Moon* (1950) with Heinlein's script; Richard Carlson and Herbert Strock's *Riders to the Stars* (1954), and Fred Wilcox's *Forbidden Planet* (1956), starring Robby the Robot), *Kubrick's 2001: A Space Odyssey* will definitely break with that. Science-fiction filmmakers explore the prose classics of the genre (e.g., Jules Verne and H.G. Wells) as source material for bolder cinematic narratives. Lang's *Metropolis* (1927) was the most emblematic movie so far. Among Well's adaptations to the cinema, big-budget Alexander Korda's *Things to Come* (1936) was the only significant attempt to accommodate the themes of British scientific romance to the medium but failed at the box-office.

The first and most successful of the space opera movie serials in this period was Frederick Stephani's *Flash Gordon* with 13—a film version of the adventures of the popular comic strip hero. Flash Gordon had his own pulp magazine. He fought against a small army of robots, created and controlled as weapons against him. It became a reference in movie representations of robots as sophisticated killing machines.

IMPACTS AND PERCEIVED RISK

As aforementioned, some experts were impacted by science fiction in general—even though we do not have survey testing for basing this statement. We do have some individual examples, some self-reports, and anecdotes that buttress that this is plausible. For example, the Hungarian-German-American physicist and inventor Leo Szilard (1898–1964), who invented the chain reaction, claimed that Well's *The World Set Free*[32] (1914) helped him conceiving the nuclear chain reaction.[33] The British biologist John Maynard Smith (1920–2004), famous for applying game theory to the study of natural selection, mentions that

> the other big influence in my childhood before I became a scientist, but obviously made me a scientist in the end, was science fiction. I read Wells, I read a man called Stapledon, who people don't read nowadays . . . wrote an extraordinary book called *Last and First Men*, in about 1932 or something, which has an atom bomb, it has a . . . the oil crisis destroying civilization, it has giant brains, it has the breeding of new human beings who will be able to build civilization and so on. And I remember reading this book and getting fascinated by genetics, because it's full of the idea of how do you change human nature by genetic means. And I guess, ultimately, that's why I became a geneticist. But there's a twist to the story, which I feel I ought to get on record, that I . . . much later, oh, quite recently, I was reading a set of short stories by Arthur C Clarke. And in this . . . there's an introduction by Clarke about his own childhood, and he describes how he read this book . . . this same book and how it blew his mind—he's a couple of years older than me, by the way—and how he . . . it made him a science fiction writer. But the fascinating thing is that he got it out of the local

public library, and so did I. . . . Science fiction had a big influence of introducing me to science.[34]

The representation of future risks of artificial intelligence varied significantly among experts. At this early moment, there was nothing concrete to fear about; there were conjectures and scenarios. However, even in this terrain, we can see the cultural differences in the United Kingdom and the United States. The succinct neutral position of Alan Turing (1912–1954) in his famous nontechnical text *Computing Machinery and Intelligence* (1950)[35] contrasts with Norbert Wiener's *The Human Use of Human Beings* (1950),[36] which provided a representation of the imagined risks thought carefully.

Turing addresses nine objections to an affirmative answer for the question, "Can machines think?" One of these objections is the future risk of machines thinking.

> The consequences of machines thinking would be too dreadful. Let us hope and believe that they cannot do so.
>
> This argument is seldom expressed quite so openly, as in the form above. But it affects most of us who think about it at all. We like to believe that Man is in some subtle way superior to the rest of creation. It is best if he can be shown to be necessarily superior, for then there is no danger of him losing his commanding position. The popularity of the theological argument is clearly connected with this feeling. It is likely to be quite strong in intellectual people, since they value the power of thinking more highly than others, and are more inclined to base their belief in the superiority of Man on this power. I do not think that this argument is succinctly substantial to require refutation. Consolation would be more appropriate: perhaps this should be sought in the transmigration of souls.[37]

As Krakauer wittily said regarding the British reception of Metropolis, "the British stayed aloof," this was the case here. Turing sounds indifferent to the perceived risk a conservative group of people had. As a matter of fact, he related this future risk representation to religious conceptions. It also seems for him that machines over-performing and overcoming of humankind was something inevitable. The discussion here is somewhat shallow, framed in terms of humankind's superior position on the planet that would be shaken. In some sense, this theological argument seems to derive from what Samuel Butler represented in *Erehwon.*[38] In the United States, there was a thorough reflection on the social consequences of artificial intelligence primarily because of Norbert Wiener (1894–1964).

At that time, Wiener was a celebrated mathematician, known for his ideas of *cybernetics*, condensed in the book with the same name (1947).[39] Some years later, he wrote a nontechnical book, *The Human Use of Human Beings*[40] (1950), to popularize the ideas of cybernetics, advocate for the automation process, and reflect on the consequences of future risks that artificial intelligence could bring. The book had a second revised edition published four years later. A quick historical note here might be useful, one of the founders of the symbolic approach to artificial intelligence, John McCarthy, coined the term artificial intelligence in the proposal for the well-known Dartmouth conference—as we showed in the introduction. He wanted to shrink from Wiener's influence and his Cybernetics School.

The Human Use of Human Beings is an example of a nonfictional work that employs a social-science-fiction style of thinking to conceive the possible consequences from high technology and especially machine intelligence. Although he knew humanity would ultimately prosper with the progress of automation with a unique role to machines in general and intelligent machines in particular, he also was a critic of the exploitation uses of technology. In this sense, he remained a pessimist at heart. In his book, he mentions a few ways the relationship with technology could be detrimental for society. Nevertheless, he was brewing his position earlier. In *Cybernetics* (1947), he posed his pessimist tone,

> There are those who hope that the good of a better understanding of man and society which is offered by this new field of work may anticipate and outweigh the incidental contribution we are making to the concentration of power. I write in 1947, and I am compelled to say that it is a very slight hope.[41]

We can see that he develops a possible technological field of application, and at the same, he envisions the consequences in the present world if it was successful. Still, in the 1950s, he communicated his representations of the risk of automation and machine intelligence buttressed on the moral and technical consequences of automation.[42] Since the end of World War II, he thought that the development of research and applied fields of electronic communication and computing machines and other automata "as a means of controlling machinery and society."[43] He conceives that, in the future, development of these communication facilities, messages between man and machines are destined to play an ever-increasing part.[44] Also, in his book *God and Golem, Inc.* (1964),[45] he returned to ethical concerns from the perspective of the creative scientist or engineer. However, he advocated emphatically for the use of robots in factories, regardless of the apparently high price these machines would cost and the relative difficulty of the tasks they would be required to do. He foresaw that the mass production of industrial robots would be feasible, and the costs will drop as new models of computing machines had been released rapidly. The mass production of computers will accompany them.

His most significative discussion starts with the scenario in which we can develop a future "chess machine" and its social consequences. Up to now, the most important representation of artificial intelligence was embodied, doubtless because of the requirement for causing physical harm. However, as soon as we can imagine an intelligent machine that can control other machines, the risk from this abstract entity becomes conceivable. Wiener thought that these classes of machines (like a chess machine) "possess some very sinister possibilities."[46] In this scenario, he brings the (science) fiction writer Edgar Allan Poe, who stared the first volume of the science-fiction pulp magazines. In one of his stories, Poe discusses a fraudulent chess playing a machine in the eighteenth century, von Kempelens's *The (Mechanical) Turk* (1770)—further acquired by Johann Maelzel depicted in Poe's story. However, Wiener proposes a genuine machine that "takes advantage of recent progress in computing machines."

The problem of combinatorial explosion[47] sets a conservative boundary on this prospective scenario, "it is hopeless to try to make a machine to play perfect chess for such a machine would require too many combinations."[48] He imagined that a machine looking two moves ahead would be satisfactory for Wiener's current knowledge of computer power. Another prominent expert involved with artificial intelligence, Claude Shannon, considered a chess machine like that.[49] He also mentions the idea of having a learning machine in this chess-playing context. Learning would be statistical learning, as soon would be developed in artificial intelligence as a scientific field. "A machine may be built either with a statistical preference for a certain sort of behavior, which nevertheless admits the possibility of other behavior, or else certain features of its behavior may be rigidly and unalterably determined," as well as rules learning. He conceived that learning machines without any hardcode constraints, such as chess-rules, could be used in a different task. It is the basis of today's transfer learning, and it has a successful application in games, more precisely Google DeepMind's AlphaGo Zero in 2017.

Regarding Wiener's book, Shannon poses that the purpose of building a chess-playing machine would be the first step toward military equipment, more precisely, to evaluate military situations and to determine the best move at any specific stage. Wiener comments on this idea,

> Let no man think that he is talking lightly. The great book of von Neumann and Morgenstern on the Theory of Games has made a profound impression on the world, and not least in Washington. When Mr. Shannon speaks of the development of military tactics, he is not talking moonshine, but is discussing a most imminent and dangerous contingency.[50]

A harmless technology as game-playing could lead to future risks. Wiener realized undesired scenarios where mathematical advances could be technically applied and help to bring human despair. As an expert, he made his position on the role of the experts and the consequences of their developments. He considered the argument that a purely intellectual or scientific job, such as a mathematician, would not be neutral, as the English mathematician G.H. Hardy has insisted in his book *A Mathematician's Apology*.[51] Wiener participated in World War II with the mathematical theory of prediction to enhance the effectiveness of anti-aircraft fire, so he saw the potential application and side effects. Practically, and in the middle of the Cold War, he positioned against engaging in more military work. He was worried about high technology *to the human use of human beings*, and not the inhumane use. Most of Wiener's colleagues and peers were offended or embarrassed by his positioning with few exceptions (e.g., Albert Einstein). So, Wiener's audience was the general educated public. He envisioned the risk of high technology (artificial intelligence included) as bringing potential social harm.

Artificial intelligence as a scientific discipline did not follow Wiener's recommendations. The field's launch was heavily funded by the militaries and for the militaries. If a systematic conception on intelligence or mind is taken apart from its systematic implementation, what remains is a highly mechanical and dehumanizing, even militaristic, outlook.

The comments of Wiener's classic book *Cybernetics* by *Le Monde's* review (1948),[52] the consequences that the commenter put forward, and Wiener's reactions to it are fascinating to understand the positions' interplay. It follows from the conception of the chess-playing machine in a scenario where it grew up and was encased in a suit of armor.[53] The reviewer-Dominican friar and philosopher named Dominique Dubarle (1907–1987)—starts with the following question: will mechanical manipulation of human reactions one day create "the best of all worlds?"[54] In a social-science-fiction thinking, Dubarle conceives the consequences of predictive statistical behavior and preferences. He imagines the implications of the optimization of the market. As we can read below, he did not fear the rise of big corporations as we fear today.

> One of the most fascinating prospects thus opened is that of the rational conduct of human affairs, and in particular . . . certain statistical regularity, such as the human phenomena of the development of opinion. Can't one imagine a machine to collect this or that type of information, as for example information on production and the market; and then to determine as a function of the average psychology of human beings, and of the quantities which it is possible to measure in a determined instance, what the most probable development of the situation might be?[55] (Dubarle, 1948)

Big technological corporations were not on the horizon yet. However, before World War II, one can see the representation of the administration of capitalists like in *Metropolis*. But Dubarle's fear was toward the State:

> Can't one even conceive a State apparatus covering all systems of political decisions, either under a regime of many states distributed over the earth, or under the apparently much more simple regime of a human government of this planet? At present nothing prevents our thinking of this. We may dream of the time when the *machine a gouverner* may come to supply whether for good or evil—the present obvious inadequacy of the brain when the latter is concerned with the customary machinery of politics. The prediction machine which determines the efficacy of anti-aircraft fire is an example of this. . . . The possibility of playing machines such as the chess-playing machine is considered to establish this. For the human processes which constitute the object of government may be assimilated to games in the sense in which von Neumann has studied them mathematically. Even though these games have an incomplete set of rules, there are other games with a very large number of players, where the data are extremely complex.[56]

The scenario Dubarle is conceiving is dystopian, in which the machines will take control over humans. The non-expert approach to a current technology deflects radically, in some moments, from an expert position. Dubarle states, "At present nothing prevents our thinking of this" whereas Wiener would say the future risk Dubarle is conceiving is not because of the technology per se,

> [It] is not frightening because of any danger that it may achieve autonomous control over humanity. It is *far too crude and imperfect to exhibit a one-thousandth part of the purposive independent behavior of the human being.*[57]

"Its real danger," Wiener added, "is . . . that such machines, though helpless by themselves, may be used by a human being or a block of human beings to increase their control over the rest of the human race"[58] (Wiener, 1954, p. 181). The threat is the uneven distribution of a technology, leading to more concentration of wealth and power in the hands of few. In the Dubarle's view, the representation of risk is the superpower surveillance state powered by artificial intelligence and portrayed in contemporary positions.

> The *machines a gouverner* will define the State as the best-informed player at each particular level; and the State is the only supreme co-ordinator of all partial decisions. These are enormous privileges; if they are acquired scientifically, they will permit the State under all circumstances to beat every player of a

> human game other than itself by offering this dilemma: either immediate ruin, or planned co-operation for outside of the very serious problems which the volume of information to be collected and to be treated rapidly still put, the problems of the stability of prediction are main beyond what we can seriously dream of controlling. . . . In comparison with this, Hobbes' Leviathan was nothing but a pleasant joke. We are running the risk nowadays of a great World State, where deliberate and conscious primitive injustice may be the only possible condition for the statistical happiness of the masses: a world worse than hell for every clear mind. Perhaps it would not be a bad idea for the teams at present creating cybernetics to add to their cadre of technicians, who have come from all horizons of science, some serious anthropologists, and perhaps a philosopher who has some curiosity as to world matters.[59]

However, Dubarle did not portray the risk of corporations sharing the lead of surveillance with the State as the social psychologist Shoshana Zuboff portrays in her *The Age of Surveillance Capitalism* (2019).[60] We will see the topic of the fears arising from dominant corporations and their nonhumane use of technology with the cyberpunk literature. The consequences here are political, arguably much more urgent, taking into account the fresh memory of the rise of Fascism and powerful states at that time. Political leaders may attempt to control their populations by means not of machines themselves but through political techniques as narrow and indifferent to human possibility as if they had been conceived mechanically. This representation of risk resembles much of the current ethical debate, especially from non-artificial intelligence experts. If the horrors of the Holocaust came to the public right after the war, this line of thinking might have been probably linked to the Holocaust and how such controlled systems could have further powered the Nazi government.

Wiener acknowledged Dubarle's attention call of the scientist to the growing military and political mechanization of the world as a great superhuman apparatus working on cybernetic principles. He attenuates Dubarle's ignorance of the state-of-art development of technology. He preserves part of his diagnosis when he argues that "even without the state machine of Pere Dubarle, we are already developing new concepts of war, of economic conflict, and of propaganda."[61] Furthermore, such conceptual development was built upon von Neumann's Theory of Games, according to Wiener, who believed government agencies were inclined to apply game theory to military and quasi-military aggressive and defensive purposes.

So, the joint call of experts and non-experts (represented by Wiener and Dubarle) raised the ethical debate and pointed to an action taking regarding the perceived risk of the development of artificial intelligence. Areas other than the pure technological ones involved become included in the discussion, "he [Dubarle] is quite right in his emphasis on the need for the anthropologist and the philosopher. [. . .] we must know as scientists what man's nature is and what his built-in purposes are."[62]

The pathfinders of risk representation were in fiction, and even a high-caliber expert as Wiener would accommodate and adapt the representation of future risk depicted in fiction, from Butler's *Erewhon* to today's technology.

> When I say that the machine's danger to society is not from the machine itself but from what man makes of it, I am really underlining the warning of Samuel Butler. In *Erewhon* he conceives machines otherwise unable to act, as conquering mankind by the use of men as the subordinate organs. Nevertheless, we must not take Butler's foresight too seriously, as in fact at his time neither he nor anyone around him could understand the true nature of the behavior of automata, and his statements are rather incisive figures of speech than scientific remarks.[63]

Wiener also explored a subtler risk from artificial intelligence than we face today, which leaves important decisions to artificial intelligence. It is even more problematic if the automated systems are based on either inflexible, literal conduct, or a black-box procedure.

> Woe to us if we let it decide our conduct, unless we have previously examined the laws of its action, and know fully that its conduct will be carried out on principles acceptable to us! . . . the machine . . . which can learn and can make decisions on the basis of its learning, will in no way be obliged to make such decisions as we should have made, or will be acceptable to us. For the man who is not aware of this, to throw the problem of his responsibility on the machine, whether it can learn or not, is to cast his responsibility to the winds, and to find it coming back seated on the whirlwind.[64]

He was already concerned with the problem of the explainability of artificial intelligence, today's pivotal topic. This kind of worry has been raised recently. It is impacting the social and economical use of artificial intelligence, especially those based on neural networks, as we will see in chapter 5.

From his communication, Wiener showed he was conscious about the adventure side of the risk-taking of developing intelligent machines. He took the consequences seriously and tried to address the real risks to the public, not the farfetched ones. Experts would have a moral obligation of educating the public, amplifying risky scenarios closer to what actually might happen, and attenuating catastrophic unrealistic scenarios. If the representation of risks propagated within society was unbalanced, the societal impacts might lead to unreflective pro-technology or anti-technology biases. Probably more than the uninformed laypeople, he wanted to make experts took a critic approach to their enterprises. As he stated, "I have said that the modern man, and especially the modern American, however much 'know-how' he may have, has very little 'know-what.'"[65]

The tone regarding his risk messages is well represented in his following prophetic passage, noticing that "prophetic means precisely the idea of telling a *future* truth,

The sense of tragedy is that the world is not a pleasant little nest made for our protection, but a vast and largely hostile environment, in which we can achieve great things only by defying the gods; and that this defiance inevitably brings its own punishment. It is a dangerous world, in which there is no security, save the somewhat negative one of humility and restrained ambitions. It is a world in which there is a condign punishment, not only for him who sins in conscious arrogance but for him whose sole crime is ignorance of the gods and the world around him. If a man with this tragic sense approaches, not fire, but another manifestation of original power, like the splitting of the atom, he will do so with fear and trembling. He will not leap in where angels fear to tread, unless he is prepared to accept the punishment of the fallen angels. Neither will he calmly transfer to the machine made in his own image the responsibility for his choice of good and evil, without continuing to accept a full responsibility for that choice . . . we shall never receive the right answers to our questions unless we ask the right questions. [. . .] The hour is very late, and the choice of good and evil knocks at our door.[66]

NOTES

1. Timothy Hornyak, *Loving the Machine: The Art and Science of Japanese Robots* (Tokyo: Kodansha International, 2006).

2. The following passage illustrates a somewhat Luddite sentiment toward machinery that are still quite alive during our days. "I also questioned them about the museum of old machines, and the cause of the apparent retrogression in all arts, sciences, and inventions. I learnt that about four hundred years previously, the state of mechanical knowledge was far beyond our own, and was advancing with prodigious rapidity, until one of the most learned professors of hypothetics wrote an extraordinary book (from which I propose to give extracts later on), proving that the machines were ultimately destined to supplant the race of man, and to become instinct with a vitality as different from, and superior to, that of animals, as animal to vegetable life. So convincing was his reasoning, or unreasoning, to this effect, that he carried the country with him" (Samuel Butler, *Erewhon: Or, Over the Range* (London: Trübner and Ballantyne, 1872). Few years earlier Butler published a newspaper article called *Darwin among the Machines* were he explore the idea of machines evolving, "What sort of creature man's next successor in the supremacy of the earth is likely to be. We have often heard this debated; but it appears to us that we are ourselves *creating*

our own successors" (Samuel Butler, "Darwing among the Machines" *Christ Church Press* [June 13, 1863]). http://nzetc.victoria.ac.nz/tm/scholarly/tei-ButFir-t1-g1-t1-g1-t4.html.

3. Frank Baum made no mistake when he represented the embodied artificial intelligence Tik-Tok. In regard of sentiments, the robot was equivalent to any other tool and it brought that fact to the time a few times. Now, why a robot would not be portrayed as being alive in a fantastic world such as Oz is unclear. The scene Dorothy, the protagonist of the Oz stories, meets the robot is quite illustrative because it has an explanatory card attached: "Dorothy stepped inside the little room to get a back view of the copper man, and in this way discovered a printed card that hung between his shoulder . . . [the card] SMITH & TINKER'S; Patent Double-Action, Extra-Responsive; Thought-Creating, Perfect-Talking; MECHANICAL MAN; Fitted with our Special Clock-Work Attachment; Thinks, Speaks, Acts, and *Does Everything but Live*; Manufactured only at our Works at Evna, Land of Ev.; All infringements will be promptly Prosecuted according to Law" (Frank Baum, *The Ozma of Oz* (Chicago: Reilly & Lee, 1907): our emphasis).

4. Mordaunt Hall, "A Mechanical Marvel," *The New York Times*, March 7, 1927. https://www.nytimes.com/1927/03/07/archives/a-technical-marvel.html?rref=collection%2Fcollection%2Fmovie-guide. Accessed June 13, 2018.

5. Siegrified Kracauer, *From Caligari to Hitler: A Psychological History of German Film* (Princeton: Princeton University Press, 1947), 150.

6. Kracauer, *From Caligari to Hitler*, 162–163.

7. Kracauer, *From Caligari to Hitler*, 32.

8. Suvin, *Positions and Presuppositions*, 101.

9. Adam Roberts, *The History of Science Fiction* (London: Palgrave Macmillan, 2016).

10. Immanuel Kant, *Critique of Practical Reason,* trans. Mary Gregor (Cambridge: Cambridge University Press, 2015), 129, 5:162.

11. Suvin, *Positions and Pressupositions*, 101.

12. Roberts, *History of Science*, 15.

13. John Rieder, *Science Fiction and the Mass Cultural Genre System* (Middletown: Wesleyan University Press, 2017).

14. Theodor Adorno and Max Horkheimer, *Dialectic of Enlightenment* (Stanford: Standford University Press, 2002).

15. See: Jefferson Pooley and Michael Socolow, "The Myth of the War of the Worlds Panic," *Slate*, October 28, 2013, https://slate.com/culture/2013/10/orson-welles-war-of-the-worlds-panic-myth-the-infamous-radio-broadcast-did-not-cause-a-nationwide-hysteria.html.

16. The typical pulp magazine had 128 pages; it was 7 inches (18 cm) wide by 10 inches (25 cm) high, and 0.5 inches (1.3 cm) thick, with ragged, untrimmed edges.

17. Roberts, *History of Science,* 263.

18. Mike Ashley, *The Time Machines: The Story of the Science-Fiction Pulp Magazines from the Beginning to 1950*, vol. 1 (Liverpool: Liverpool University Press, 2000).

19. Roberts, *History of Science,* 271.
20. Everett Bleiler, *Science Fiction: The Gernsback Years* (Kent, Ohio, and London: The Kent State University Press, 1998).
21. Bleiler, *Science Fiction.*
22. Bleiler, *Science Fiction*, xxix.
23. Asimov, "Social Science Fiction," 169.
24. Bleiler, *Science Fiction*, xiv.
25. Bleiler, *Science Fiction*, xiv.
26. Bleiler, *Science Fiction*, xvii.
27. Marvin Minsky, interview by Christopher Sykes, January 31, 2011, https://www.webofstories.com/play/marvin.minsky/52.
28. Marvin Minsky, interview by Christopher Sykes, January 31, 2011, https://www.webofstories.com/play/marvin.minsky/52.
29. Bleiler, *Science Fiction*, vii.
30. Bleiler, *Science Fiction.*
31. Roberts, *History of Science*, 254.
32. In this story, Wells conceive a scenario clean, efficient atomic engines in England during the 1950s. A world war breaks out, in which atomic bombs wipe out the world's great cities.
33. Rhodes, *Making of Atomic Bomb.*
34. John Maynard Smith, interview by Richard Dawkins, April 1997, https://www.webofstories.com/play/john.maynard.smith/1.
35. This is the paper in which Turing proposed his imitation game known today as the Turing Test. Alan Turing, "Computing Machinery and Intelligence," *Mind* 49, no. 236 (1950), 433–460.
36. Norbert Wiener, *The Human Uses of Human Beings*, 2nd ed. (London: Free Association Books, 1989 [1954]). First edition was published in 1950.
37. Turing, "Computing Machinery."
38. Butler, *Erewhon.*
39. Norbert Wiener, *Cybernetics: or, Control and Communication in the Animal and the Machine* (Paris: Technology Press, 1948).
40. Wiener, *The Human Uses.*
41. Wiener, *Cybernetics*, 39.
42. Norbert Wiener, "Some Moral and Technical Consequences of Automation," *Science* 131 (1960), 1355–1358.
43. Wiener, *The Human Uses*, 15.
44. Wiener, *The Human Uses*, 16.
45. Norbert Wiener, *God & Golem, Inc.: A Comment on Certain Points Where Cybernetics Impinges on Religion* (Cambridge: MIT Press, 1964).
46. Wiener, *The Human Uses*, 175.
47. As generally defined, a combinatorial explosion means the rapid growth of complexity of a problem due to how the number of possibilities of the problem is affected by the input, constraints, and bounds of the problem.
48. Wiener, *The Human Uses*, 175.

49. "I am willing to accept Shannon's conjecture that such a machine would play chess of a high amateur level and even possibly of a master level." Wiener, *The Human Uses*, 16.

50. Wiener, *The Human Uses*, 178.

51. Geofrey Harold Hardy, *A Mathematician's Apology* (Cambridge: Cambridge University Press, 2004).

52. Dominique Dubarle, "Une Nouvelle Science: La Cybernétique—Vers La Machine à Gouverner?," *Le Monde*, December 28, 1948, http://sniadecki.wordpress.com/2013/03/08/dubarle-machine/.

53. Wiener, *The Human Uses*, 178.

54. Origianally in French: *La manipulation mécanique des réactions humaines créera-t-elle un jour «le meilleur des mondes»?*

55. Dubarle, "Une Nouvelle Science."

56. Dubarle, "Une Nouvelle Science." This piece of text belongs to Wiener's translation, *The Human Uses*.

The original French is available in: http://archive.wikiwix.com/cache/index2.php?url=http%3A%2F%2Fsniadecki.wordpress.com%2F2013%2F03%2F08%2Fdubarle-machine%2F.

57. Wiener, *The Human Uses*, 180–181—our emphasis.

58. Wiener, *The Human Uses*, 181.

59. Dubarle, "Une Nouvelle Science."

60. Shoshana Zuboff, *The Age of Surveillance Capitalism: The Fight for a Human Future at the New Frontier of Power* (New York: PublicAffairs, 2019).

61. Wiener, *The Human Uses*, 181.

62. Wiener, *The Human Uses*, 184.

63. Wiener, *The Human Uses*, 182–183.

64. Wiener, *The Human Uses*, 185.

65. Wiener, *The Human Uses*, 184.

66. Wiener, *The Human Uses*, 184ff.

Chapter 3

We Have Something to Do

Defining Artificial Intelligence amid the Scientific Space Odyssey in the 1960s and 1970s

After World War II, the United States and the Union of Soviet Socialist Republics (URSS) were disputing to bring the more Third World countries they could to the side of either capitalism or socialism. In the middle of this political race in 1957, the URSS launched Sputnik, a satellite, into space. One month later, they launched Sputnik 2 with a dog inside. Laika, the dog, survived to the spatial race, and the URSS proved to the world that their technological advancements were no empty political propaganda. The entire world stopped to see the rockets. They crossed the American sky leaving behind the fear of being overcome by the Soviet Union. Indeed, URSS left far away behind the United States in the technological dispute at that moment. However, the launch of Sputnik frightened the United States by showing that they were close to having a nuclear war, once they already had the technological background. Sputnik was the 9/11 of that time.

After that, it started to be inquired in the United States how they could have been overcome technologically by the URSS. This event sparked the national feeling, usually called as "Sputnik Crisis." Different voices ranging from intellectuals to politicians kept saying that the United States had been at risk and their security at stake. Although any real threat was not even close to happening, the "Sputnik rhetoric" was very frightening to Americans. In the meantime, some rocket launches' failures justified their fears (e.g., the Vanguard TV2, the Vanguard TV3, and the Vanguard TV3B). To bridge this technological gap, the United States redirected scientific and technological targets. Instead of focusing on producing gadgets that settled the "American life dream," they focused on training children in sciences and later convert them into real scientists. The "baby boom" and the "scientist boom" came together in the turn of the 1950s to the 1960s, as well as the reform of the educational structure. Related or not to this boost, in 1969, the United States could send man to the space.

Those were relevant aspects of the sociohistorical context in the passage from the 1950s to the 1960s that would impact science, technology, and fiction. The spatial race was the background that illuminated both the cultural representations on artificial intelligence (AI) and the real AI developments. The space race and the robots attached to space scenario had boosted the scientific developments in general. In the popular culture, science fiction had been imagining space travels and humans flying to the moon for a long time. Nevertheless, a new concept in this period inspired new scientific ideas, such as the "cyborg." Manfred Clynes and Nathan Kline presented this idea in *Cyborgs and Space* (1960), an article for the journal *Astronautics.* The word "cyborg" stands for Cybernetics' Organism, a conception explicitly buttressed on Wiener's cybernetics. Clynes and Kline proposed that,

> Altering man's bodily functions to meet the requirements of extraterrestrial environments would be more logical than providing an earthly environment for him in space. . . . Artifact-organism systems which would extend man's unconscious, self-regulatory controls are one possibility.[1]

Although very innovative, the cyborg-cybernetics idea kept silent within the culture at that period. An important reason for such latency is the fact that cybernetics did not turn out to be the dominant view of AI in the 1960s. The same happened with the neural networks approach. There was an institutional selection that made it happen, namely the Advanced Research Projects Agency (ARPA), currently named as Defense ARPA (DARPA). This agency was created in February 1958 as an institution to look out the societal impact of the fears caused by the Soviet launching of Sputnik a year before. The selected awardees by ARPA belonged to the symbolic approach; they were John McCarthy, Marvin Minsky, Herbert Simon, and Allen Newell.

The political and institutional environment of this period, the different approaches to AI, and the popular representation of fears and anxiety about what computers could do are brilliantly tied together in a satyrical text by Louis Fein called *The Artificial Intelligentsia* published in the *IEEE Spectrum* (1964). With the technical supervision of Nils Nilsson, Dr. Fein established the first Computer Science Department in the United States at Purdue University in 1962. As a Palo Alto consultant, he also advocated for the research on computers not driven by military purposes in *Computer-oriented peace research* (1963).[2] A year after, in *The Artificial Intelligentsia* paper, he asked the audience to imagine the following:

> I will ask you to imagine that a Federal agency has sent a "request to bid" on some work in artificial intelligence to four previously qualified bidders, each institution representing a contending school of thought. Each bidder writes a proposal. All proposals are evaluated by a competent and skeptical technical expert who represents and advises the agency. This review consists of six sections: a fictitious request to bid, four proposals, and technical evaluation.[3]

Dandylines Enterprises, Search Ltd., and OPTYmistica stand for neural networks, symbolic AI, and cybernetics, respectively. Search Ltd. was the preferred solution because it proposed practical applications, simulations, and open to inspection. Dandylines had a too academic approach, as will see, and OPTYmistica was disregarded given the obscurantism of its proposals by the evaluator, the *Pessimyths*—"An evaluation of OPTYmistica will not be made; its maladaptive proposal speaks for itself."[4] The Edisons of the companies, as the Pessimyths reported, were "enthusiastic and optimistic. They have promised such things as champion chess masters, music composers, infallible and quick reference librarians, spoken language translators, and good mathematicians."[5] The report concludes that "They have failed to deliver," and recommends "against contract awards because the whole field is in the research phase."[6] J.R. Bubble Piercer, the president of the Pessimyths, signed the report.

Although the forming bubble had been fully recognized not only by Louis Fein but also by the philosopher Hubert Dreyfus at the same, it was pierced almost a decade later, when Sir James Lighthill, the distinguished British applied mathematician played the role as the "J.R. Bubble Piercer." Despite that, we should acknowledge that AI people had reasons to be enthusiastic, since their scientific enterprise had a sequence of achievements such as theorem provers, solving calculus problems, blocks manipulation, a chatbot (Eliza), checkers playing, even a prospect of chess was already in the horizon. Ironically, the hard problems for regular people were solvable with computers, whereas what we think as easy activities involving emotions and commonsense turned out to be extremely difficult. This gap allowed some non-experts, but still close observers, to criticize the field harshly. The most striking example was Dreyfus criticism on the idea that experts were making progress toward developing real AI, "According to this definition [of progress], the first man to climb a tree could claim tangible progress toward a flight to the moon."[7]

Although the overpriced general expectations of AI came to a correction, historically dramatized in the Lighthill's Report and the BBC public debate broadcast, the representation of AI's future (including its risks) managed to be root deeper in culture. It could not be different given the electric environment this period was experiencing, chiefly in the United States. It is illustrative to see how Louis Fein started his paper,

> Thinking machines, translating machines, learning machines, self-organizing systems, conditional probability machines, teaching machines, adaptive machines, self-reproducing machines, man-machine symbiosis, intelligence amplifiers, artificial intelligence, heuristics, list processing, pattern recognition, simulation of cognitive processes, brain modeling, cybernetics, synnoetics, concept formation, bionics, and scores of other such terms may evoke awe, despair, confusion, charges of fraud and charlatanry, hope for the world, fear of abnegation of man's control of his own affairs, hope for a leisure society, fear of unemployment, etc. *What's the fuss about?*[8]

All these bits charged by this overall project of machine intelligence boomed in a short period. In this regard, as we saw in the introduction of the book, the term "artificial intelligence" served *as a magnet term* for all these charged bits as well as all these sentiments toward those mentioned by Fein.

During this period, Stanley Kubrick and Arthur C. Clarke's *2001: A Space Odyssey* amalgamated science-fiction representations, the state-of-the-art technology and its unfoldings, and the general anxieties and hopes regarding space's project into a single popular project. In many aspects, the movie became the paradigmatic representation of AI and the risks we might face. Still, regarding computers and AI in the context of the 1960s, a regular person during the 1960s both loved computers and was scared of them, as one of the founding fathers of AI, Marvin Minsky, remember. He underscores that "there are no benevolent superintelligent computers in film."[9] The science-fiction representations of the risk of intelligent machines were pervasive within, at least, American society. With the advancement in computer hardware and software during this period, David Kuck, a Computer Pioneer awardee[10] highlights that "the application that jumped to the forefront of the minds of computer scientists and laymen as soon as modern computers were invented was AI, under various definitions."[11]

In the cultural front during this period, besides Arthur C. Clarke's creation, we still have some episodes of celebrated TV Shows in which machine intelligence appeared, such as Rod Sterling's *The Twilight Zone, Doctor Who, Lost in Space*, and *Star Trek*. In Japan, the pop culture character Doraemon, a robotic cat, appeared as an *animé* (1969) and TV series "*anim*" (1973, 1979) for a very young audience.

Philip K. Dick wrote significant parts of his work during this period. The most notable for the cultural representation of the risks of artificial was *Do Androids Dream of Electric Sheep?* (1968), which served as the baseline story for the second most excellent AI movie in the next period, Ridley Scott's *Blade Runner* (1982). Dick also portrayed a series of disembodied AIs in *Vulcan's Hammer* (1960), his last pulp story, and in *A Maze of Death* (1970). Another emblematic AI was the character *Deep Thought* (1978) in the comedy radio series *The Hitchhiker's Guide to the Galaxy* by Douglas Adams. At the end of this period, movies that depicted intelligent machines in a plan to take over the world still existed; see, for instance, the *Gunslinger*, a character in Richard Heffron's *Westworld* (1973) and *Futureworld* (1976). In 1977, the *Star Wars* brought some popular intelligent machines, such as C3PO, R2D2, as servants, and general *droids*.

HAL INTO SPACE: THE REALISTIC REPRESENTATION OF AI AND ITS RISKS

"*2001* transcends the label of 'science fiction movie' and captures many of the central metaphors of *our* time, telling us much about society and its aspirations. [. . .] Many people have been deeply affected by the film, among them several contributors who reflect here about its influence on their own careers and on computer science in general."[12] These are the open remarks of David Stork, an expert on pattern recognition, in a book about Hal 9000's legacy. Whatever the real impact the Kubrick's *2001* exerted on the computer scientists of the period, the new generation of researchers that came right after will take the AI Hal 9000 as a technical milestone to be achieved. A reasonable argument is a comparison with space travel achievements. Since the movie's debut, we had humans landed on the Moon, robots exploring the surface of Mars, and a probe that reached Pluto. Why could we not develop AI at the same pace?

The importance of Kubrick's movie and, consequently, the role of Hal 9000 seemed to increase in the past decades. Although we lack a proper study on that, the American Film Institute's rankings may help us to have a glimpse of this phenomenon. According to the Institute's ranking of the best movie of all time, Kubrick's *2001: A Space Odyssey* was considered the twenty-second best movie of all time in 1998, but it moved to the fifteenth place in 2007. Similarly, *Blade Runner*, which was absent from the 1998s ranking, appeared in 2007 in ninety-seventh position.[13] Still considering the American Film Institute information, *HAL 9000* ranks a thirteenth best villain. The cyborg *T-800*, played by Arnold Schwarzenegger in James Cameron's *The Terminator* (1984), ranks in twenty-second position. However, *T-800* appeared as the forty-eighth best hero in the *Terminator 2* (1991) movie. No other AI ranks either in the villain or in the heroes' ranking.

We may argue that science-fiction stories in the eyes of the public used to be depreciating the value of the scientist endeavor as something exotic, dangerous, ran by eccentric people, at least until 1968. Such stigmatization was a side effect. Some scientists chose to take these cinematic representations as jokes, whereas others preferred to fulminate against their implausibility.

The association of intelligence and machines was not positive. Minsky also remembered in an interview about *Hal 9000* that "up to about 1985, IBM had a rule against employees stating that a machine could be smart, or had artificial intelligence."[14] Science-fiction narratives were also really hard to implement; we have been in contact with more science-fiction productions because of the advances in computer graphics. The production of *2001: A Space Odyssey* (1968) fought against the previous cliché and impossibilities trends. Its fictional world was stemming from a careful expert assessed technologies dominate science-fiction cinema in the last decades; for example, George Lucas's *Star Wars* (1977). It shaped the way we have imagined the future in space and even what AI might be. As Alan Kay said, the best way to predict the future is by inventing it.

2001: A Space Odyssey was a joint enterprise. Some of the most prominent experts of the time in their research field were involved in the production of the movies. When it comes to AI, Marvin Minsky and I.J. Good were deeply involved—although the last is mentioned only in the preproduction of the film. Kubrick got into contact with Minsky via Arthur C. Clarke,[15] and he was interested in the computer graphics done at MIT. As an expert, Minsky was very impressed with how carefully Kubrick and Clark dealt with science in their movie, instead of appealing to "technobabble"[16] (Stork, 1997). Hal 9000 created a single application for different researches going on in AI.

> Hal presents to researchers a powerful cinematic representation of AI precisely because he simultaneously demonstrates general intelligence while keeping visible the AI sub-problems, roughly corresponding to different sub-fields within AI. Thus researchers can easily recognize AI specialties in Hal's individual capabilities, making Hal plausible, while seeing the individual capabilities integrated into a general intelligence, making Hal compelling.[17]

Minsky, as an expert consultant, did not know much about the role of Hal 9000. Still, he helped with technical details, such as the suggestion for the name "Hal" standing for *H*euristically programmed *AL*gorithmic computer, the idea that the computer voice would not be so smooth as a human speaker, and so on. He had no role in conceiving the side effects provoked by Hal. When it comes to I.J. Good, we also do not know the extent of his influence. In Clarke's book, he wrote this kind of history of the future that "in the 1980s, Minsky and Good had shown how neural networks could be generated automatically—self replicated—in accordance with any arbitrary learning program."[18] However, Good had conveyed a risk perception of AI linked to the intelligence explosion (1965), a topic explored by "future futurists," such as Vernor Vinge, Ray Kurzweil, and Nick Bostrom. The risk from Hal 9000 had a different origin. Clarke furnished a hint in the book: Hal could not keep secrets without showing malfunctioning. He knew the truth of the mission, but he had to keep the secret.

> He was only aware of the conflict that was slowly destroying his integrity—the conflict between truth, and concealment of truth. He had begun to make mistakes, although, like a *neurotic* who could not observe his own symptoms, he would have denied it.[19]

In that sense, Clarke shows how to conceive an imaginary risk scenario. The consequences of having a system that has to integrate its knowledge and behavior would be unable to deal with this typical social requirement. Once the computer can represent its inner states symbolically, one will be able to describe them with psychoanalytical vocabularies, such as "neurotic." From here, it is possible to hypostasize a side effect AI that has never been imagined yet: the emergence of a superintelligence without the concept *of social roles*[20] capable of dealing with different audiences. This means that at odds with humans, a computer could not manipulate his facade according to whom it is interacting. Although the machine could integrate all the information, it could not discern in which circumstances it should use them. By trying to solve that problem, the binary system of Hal 9000 would suffer from collapses.

It is puzzling to define the role of Stanley Kubrick in terms of risk representation of AI in his movie. He was not, of course, shooting a film *about* AI, it was a movie *with* AI. In the following years, he tried to develop a movie only about AI based on the writer Brian Aldiss's short story "*Supertoys Last All Summer Long*" (1969). The movie will be accomplished after Kubrick's death, by Steven Spielberg, released as *A.I. Artificial Intelligence (2001).* Several mundane ideas regarding of how Hal shows itself for the audience came from him. Up to these days, the scenarios in which AI may pose a risk are not realistically depicted. Thereby, they are dull scenarios tending to cliché or empty ideas. In this sense, Kubrick played a pivotal role in the formation of risk representation of AI because he helped to form it by drawing counterfactual scenarios by propagating it through this movie. This was not something taken-for-granted. He knew he would need the general experts' approval to fall into the traditional scientist's rejection of science-fiction narratives. For that, he tried to prepare his audience not only with technical precision but also with a proactive approach. For example, in Michael Benson's precise book on the making of *2001*, he mentions the following:

> Early in preproduction, Kubrick had conceived of a documentary prologue, designed to protect 2001 from the kind of misunderstandings engendered by the then prevalent Buck Rogers and Little Green Men school of science fiction.

> Featuring eminent scientists such as Freeman Dyson, Margaret Mead, B. F. Skinner, and eighteen other leading researchers discussing such subjects as extraterrestrial life, space travel, and communication between alien civilizations, the prologue was entrusted to Roger Caras[21] and was actually shot, on black-and-white 35-millimeter, throughout 1965.[22]

In sum, with *2001*, Kubrick and Clarke represented a realistic future risk, but also attenuate other more unrealistic risks. The point is that they amplified the representation of risk regarding AI, which had its side effects on a new generation of experts, and a wider audience for AI—now, with much higher bar.

Moving to how the scenario regarding intelligent machines developed, we have a relaxed attitude toward them in science-fiction since the 1950s. Robots were being "tamed." Industrial automation had made progress, and the first computer-controlled industrial robot arms appeared on production lines in 1960. In the same year, the (defunct) American aerospace company, Hughes Aircraft Corporation, built a mobile robot designed to work in dangerous areas. At the end of the decade, the Stanford Research Institute's deployed Shakey, the first general-purpose mobile. Whether or not in parallel to the actual developments in robotics, the great fiction stories on robots in the 1970s will call attention to their (imagined) subjective element. Philip K. Dick, one of the most important science-fiction writers, will be a pioneer in this field. Some other notable contributions are the aforementioned Brian Aldiss's *Supertoys* (1969), Stanislaw Lem's collected robotic fables in *The Cyberiad* (1974) and *Mortal Engines* (1977), Barrington J. Bayley's *The Soul of the Robot* (1974), and the still ongoing development of the Asimov's robot stories, especially *The Bicentennial Man*' (1976) that explores the limits of the definition of the social interpretation of humanity.

Although the importance of these social-science fiction, their impact during this period was not widespread as Clarke and Kubrick's *2001*. Later, in the same way, Dick's *Do Androids Dream of Electric Sheep?* (1968) also had played an important role in popular culture regarding the future and the side of effects of AI mostly because of Scott's *Blade Runner*. The book was published in the same year as Clarke's *Space Odyssey*, but it brought robots instead of the disembodied version of AI. If Hal 9000 became the paradigmatic representation of the system-like abstract AI, Dick's *Replicants* (how the robots or androids are called) might be considered the paradigmatic image of human-like robots. In a post-apocalyptic San Francisco (note the in *Blade Runner's* setting is in Los Angeles), six androids escape and have to be killed. The one who had been charged for this task was Rick Deckard, a bounty hunter. The story explores the intricacies of empathetical relations with machines. The androids have no genuine emotions, despite that one of them managed to seduce Deckard for ending with his mission. This way, there isan emotional risk, a "deception threat," because we may believe that a machine has the same relationship toward us as humans have—when, in fact, they do not. For unpuzzling this relationship, Deckard has an empathy test to check if a robot is a human or not. This point is fascinating because it conceives a scenario in which machines had conquered the Turing test. Consequently, a robot becomes indistinguishable from a human being. The inverse rationale can be applied. A human being is indistinguishable from a robot. In this sense, we must have a humanity test, which was thought of as an empathy test. Such a test brings its side effects. If an advanced machine performs well in the test, the tester may be bewildered about his condition, starting to suspect that the examiner him/herself is a robot! That is precisely what happened to Deckard. This scenario was explored more recently in Alex Garland's movie *Ex-Machina* (2014) when the protagonist cuts himself to be assured that he was a human, not a robot.

AI IS MOVING ON: FIRST DEVELOPMENTS OF WHAT WE CURRENTLY CALL AI

In a stock market metaphor, the extrinsic value of AI was many times higher than its intrinsic value. However, the latter was not null. Quick breakthroughs in AI research brought the impression the challenge of devising a true AI was on the horizon - at least from the experts' perspectives. External observers had quickly devolved some skepticism regarding the promises of AI research. The shrewdest observer was Hubert Dreyfus.

He was summoned by the RAND Corporation in Santa Monica at that time to join to the AI projects as a consultant studying AI research. The RAND Corporation wanted a philosopher in the group, and Dreyfus was a philosophy professor at UC Berkeley. However, he began his career at MIT, so he saw how the AI enterprise played out there with Marvin Minsky. He spent the summer of 1961 at RAND, and shortly after the summer, he wrote RAND a paper titled *Alchemy and Artificial Intelligence*, in which, among other things, he concluded that the ultimate goals of AI research were as unachievable as were those of alchemy. The tension would escalate into a dramatic event. The so-called *Lighthill Report* would bring it to an end. Dreyfus's positioning was made probably too soon. He developed an opposition in principle to what he saw in AI research at the moment. Even though most of the problems could be called "toy" problems, as Seymour Papert put it, yet they were the initial research in the field.[23]

There has been some progress in pattern recognition research, notably with the development of the early neuron networks approach. The psychologist

Frank Rosenblatt (1928–1969) began working on neural networks in 1957 and published a paper on the subject a year after in the journal *Psychology Review.*[24] Rosenblatt called these networks as *perceptrons* as potential models of human learning, cognition, and memory. Rosenblatt's initial plan was to develop hardware rather than software; he favored a brain-style approach to neuron networks, instead of simulating them. One of the main applications was the recognition of characters, a problem that Oliver Selfridge and Ulrich Neisser were tackling at the same time by reporting their work in *Scientific American* in 1960. According to Nilsson, they employed successfully "image processing, feature extraction, and learned probability values in hand-printed character recognition."[25] Rosenblatt's approach (the connectionist approach) as the chiefly alternative to symbol-processing methods. He further developed his ideas in *Principles of Neurodynamics.*[26] Rosenblatt got a lot of publicity. The work on perceptrons set the stage for contending views represented by Rosenblatt and Marvin Minsky, who wanted to establish its limitations—a task that Minsky felt was a sort of social service they could perform for the AI community.

The ARPA, later DARPA, came as an institutional answer for the social impact caused by the threat represented by Sputnik. It became the main institutional player to fund the research on AI during this period. There were some approaches in this broad area, but only the symbolic approach became the paradigm. ARPA backed symbolic AI explicitly, but it refused money to neural networks research[27] because it seemed more classic scholar work and less applied research.

Supported by ARPA funding, the initial leading core of symbolic AI—a reduced group of researchers and their students, working at a few prestigious centers such as MIT (Minsky's group), Carnegie-Mellon University, Stanford University (McCarthy's group), and SRI—had a privileged access to economic and (the then so scarce) computing resources, and consolidated their professional and organizational network. ARPA's policy favored resource concentration at a few centers of excellence, and the selection of projects was based neither on peer review nor on equalizing principles for research money distribution, but on the agency's judgment about the best researchers working on the best projects for the agency's military goals. According to Paul Edwards:

> Funding from ARPA was virtually automatic; Licklider simply asked McCarthy what he wanted and then gave it to him, a procedure unthinkable for most other government agencies. Licklider remembered that 'it seemed obvious to me that he should have a laboratory supported by ARPA. [. . .] . . . So I wrote him a contract at that time.[28]

An early development was Allen Newell, John C. Shaw, and Herbert Simon's *General Problem Solver* (1959), which dealt with subproblems of a bigger problem applying recursion to solve them. A recursive program means a program that calls itself during its running. The *General Problem Solver* was able to compute the "difference" between a problem to be solved and what was already known or given. Newell and Simon's idea is that humans use such a hierarchical organizational approach to solve problems.

John McCarthy and his students worked with knowledge representation and reasoning, working with symbolic logic. One of McCarthy's main contributions was developing a high-level programming language called List Processing (LISP), in 1958. The goal of LISP was representing everything as a list data structure—even the program itself, and treat computation as the evaluation of mathematical functions. LISP became the reference language for functional programming—one of the paradigms of programming languages. In the early 1970s, Alain Colmerauer and Robert Kowalsky made available the logic programming or Prolog, which is a language based on first-order logic (aka. predicate logic). It is also a reference for the logic programming paradigm.

Natural language processing started early. In 1957, the American linguist Noam Chomsky published a ground-breaking book titled *Syntactic Structures* introducing transformational generative grammar, a set of grammatical rules that could be used for generating the "legal" sentences of a language. Following Chomsky's work, it was also introduced with that phrase markers, or P-markers, representing the deep structure (the abstract representation of the syntactic structure) of a sentence is generated by applying phrase structure rules. This approach is called *constituency grammars*—as opposed to the *dependency grammars*. Terry Winograd developed a LISP program, SHRDLU, in which the user could move objects in a virtual world via a dialog. There were also speech recognition programs. HARPY (1976), for instance, was a system stemmed from the combination and improvement of two earlier systems, HEARSAY-I (1975) and DRAGON (1973), all developed by Carnegie-Mellon University. Later, Roger Schank and Robert Abelson, in *Scripts, Plans, Goals, and Understanding*,[29] had presented some natural language understanding based on the ideas of scripts and frames. But the Holy Grail application of this period was machine translation, which turned out to be a problem. Many researchers were skeptical about automatic translation, Norbert Wiener included. He confessed the following, "I frankly am afraid the boundaries of words in different languages are too vague, and the emotional and international connotations are too extensive to make any quasi-mechanical translation scheme very hopeful."[30] Lighthill attacked the machine translation difficulties in his report and during the BBC broadcast.

Because of the Vietnam War, the U.S. Congress passed the "Mansfield Amendment" to the Defense Procurement Authorization Act of 1970 (Public Law 91-121), which restricted the support to basic research to those with a direct contribution to military operations. In 1972, ARPA became DARPA, and the funding for AI almost faded out.

WHEN DISPERSED TECHNOLOGIES WERE NAMED: DISCUSSION AMONG THE "ARTIFICIAL INTELLIGENTSIA"

In the early 1970s, UK's Science Research Council (SRC) was increasingly receiving applications for research support in the "broad field with mathematical, engineering and biological aspects which often goes under the general description Artificial Intelligence," as Sir James Lighthill defined it in his report.[31] The head of the SRC, Brian Flowers (1924–2010), had commissioned him in 1972 to give an unbiased view of the state of AI research primarily in the United Kingdom. Sir Lighthill, as an applied mathematician, known for his pioneering work in the field of aeroacoustics, accepted, and did it under the name of "Artificial Intelligence: A General Survey." A BBC television program on the "Controversy" series followed the report in 1973 under the provoking name "The general-purpose robot is a mirage."

Within his report, Lighthill organized the field in three categories of activities: the advanced s a mean of automation of human activities, the theoretical investigations related to neurobiology and psychology, and a bridge category based on the construction of robots. His categorization was not strikingly controversial, as we can see on Professor Sutherland's comments on the report, "no one is likely to quarrel with Lighthill's classification of work in AI as falling roughly into three areas" or as showed by the BBC program, which was organized mirroring the structure of Lighthill's Report: Lighthill as the discussant, and three experts representing each one of the activities, Professor Donald Michie (1923–2007) for the automation, Professor Richard Gregory (1923–2010) for the neurobiology and psychology team, and Professor John McCarthy representing the bridge activity, who was seated between Michie and Gregory. Now, Lighthill's claim about AI as a field was that its "fundamental coherence . . . must depend on one's opinion on whether the arguments for this Bridge Activity [. . .] are sound enough for it to be regarded as a necessary concomitant to, and the link between, the rather different and rather easily defensible activities."[32] While he encouraged the first two categories of activities, his critiques on the bridge activity were discouraging. It raised doubts about whether the whole concept of AI as a unified field of research is a valid one.

The problem with this bridge activity for Lighthill can be summarized in a single word: "naivety." The title of the BBC program explicitly shows that. In the report, I think Lighthill tries to persuade the audience to position himself against this nonrealistic optimism that the researchers in the bridge front held. His critique was not against the too optimistic positions from the side of AI but against all of them. In the British sense of humor, he calls wittily but mockingly the representatives of AI as the "Artificial Intelligentsia." For instance, he says,

> When able and respected scientists write in letters to the present author that AI, the primary goal of computing science, represents another step in the general process of evolution; that possibilities in the nineteen-eighties include an all-purpose intelligence on a human-scale knowledge base; that awe-inspiring possibilities suggest themselves based on machine intelligence exceeding human intelligence by the year 2000; when such predictions are made in 1972 one may be wise to compare the predictions of the past against performance as well as considering prospects for the realization of today's predictions in the future.[33]

Lighthill astutely shows the clash between prior expectations and the immense difficulties of achieving some tasks at that time, for example, "achieving good hand-eye co-ordination[34] in quite simple problem situations."[35] Lighthill also says that since the beginning of this enterprise of building robots in a general-purpose way, some able research workers were aware that it could be "a precarious or even crazy enterprise." But they were "attracted to participate in such a long-shot or even shot in the dark activity."[36] In sum, as he emphasizes, "some workers in the field freely admit that originally they had very naive ideas about the potentialities of intelligent robots but claim to recognize now what sort of research is realistic."[37]

These broken expectations caused by facing reality is something that exists in the AI's horizon since its foundation. During the emblematic Dartmouth Conference, McCarthy exposes his disappointment later: "I simply measured the distance between what I had hoped to accomplish and what we did accomplish, and it was pretty large."[38] In March 1989, in an interview for the program *Thinking Allowed*, McCarthy expresses the same feeling again, "the field has made somewhat less progress than I hoped, although I didn't have any definite opinion as to how fast it would progress."[39] These declarations of McCarthy suggested that he had his feet on the ground, and he was pretty much a pragmatist. During the BBC debate, he told how he picked "Artificial Intelligence" as the name of his enterprise,

> I invented the term AI because we had to do something, and we were trying to get money for a summer study in 1956. I had previous bad experience [. . .] in 1952 when Claude Shannon and I decided to collect a batch of researches which we hoped would contribute to launching this field. Shannon thought that AI was too flashy a term and might attract unfavorable notice. So we agreed to call it

> "automata studies," and I was disappointed when the papers we received were about automata. Very few of them had anything to do with the goal that at least I was interested. So, I decided not to fly any false flags anymore; it is a study at the long-term goal of achieving human-level intelligence. Since that time, many people have quarreled with the term, but have ended up using it. Newell and Simon, the group at Carnegie Mellon University, tried to use "complex information processing," which is undoubtedly a very neutral term. Still, the problem was that it didn't identify their field because everyone would say, "well, my information is complex, I don't see what's special about you.[40]

The term was not just a happy fashion drew from a sci-fi book, McCarthy put much thought on that, and he positioned himself firmly against other candidates at that moment. We have just seen how he wanted to be different from the automata theory. Still, he also wanted to distinguish the discipline from an older and important area, the cybernetics. He has declared in the review he made for the book *The Question of Artificial Intelligence*[41] that he kept the term AI "to escape association with 'cybernetics.'" Because "Its concentration on analog feedback seemed misguided, and I wished to avoid having either to accept Norbert Wiener as a guru or having to argue with him."[42] One cannot underestimate how successful McCarthy has been. Then Complex Systems was a solid intellectual enterprise, which had linked to the vast, very influential scholars such as Ludwig von Bertalanffy (1901–1972), Ross Ashby (1903–1972), and Norbert Wiener (1894–1964) himself. Even how he did conceive intelligence in terms of goal achievement as being the "computational part of the ability to achieve goals in the world."[43] Although problematic in the sense of the questions regarding the goals, for instance, "how we make these goals in the first place?," this definition is still strict.

His insistence on making the discipline an autonomous scientific field is worthy of mentioning. For example, in the written reply to the *Lighthill Report*, he emphasized that the "Bridge Category" is not subsidiary of the others, but "studying the structure of information and the structure of problem-solving processes independently of applications and independently of its realizations in animals or humans."[44] Although McCarthy was optimist with AI, in the long run, he was not a futurist. In his oral history in an interview by Nils Nilsson in September 2007, McCarthy is very skeptical of the short-term success of AI, "short-term success has eluded people," in the sense that "human-level artificial intelligence is quite likely to be a long-term proposition."[45] He illustrates it better by positioning against the extreme optimism of Ray Kurzweil, the American futurist, inventor, and entrepreneur, saying that the current enthusiasm for short-term success is just a promotion by Kurzweil.

> He wrote a book called "The Singularity is Near" (Kurzweil, 2005), I read his chapter on Artificial Intelligence, and as far I am concerned, he has not provided any sufficient basis for his short-term optimism. The chapter is not technical . . . it merely consists of his references to various demos.[46]

Despite McCarthy's pragmatism, Lighthill made promoted a *tabula rasa* against AI, the idea to have general-purpose robots. For him, because the problem of "combinatorial explosion," which implied that many of AI's most successful algorithms would grind to a halt on real-world issues and because of that they were only suitable for solving "toy" versions of them—like in games—the whole enterprise would be "a mirage." While this movement of AI developments toward software was an accurate prediction, he considered it less valuable than a full-fledged general robot. Contrariwise, for McCarthy's understanding, games would have a critical experimental role for the discipline, in particular, the game chess, which he mentions to be, according to an unknown Russian computer scientist, "the Drosophila of Artificial Intelligence."

Facing his diagnosis, Lighthill offers an explanation of the intelligentsia's driving forces as being human imagination and fiction:

> [W]e have to remember the long-standing captivation of the human imagination by the very concept [of building robots], as shown by its continual prominence in literature, from medieval fantasies of the Homunculus through Mary Shelley's Frankenstein to modern science fiction. To what extent may scientists consider themselves in duty bound to minister to the public's general penchant for robots by building the best they can?[47]

"Duty" is a fascinating word in the idea of developing intelligent robots. The most popular argument is the need for developing AI because human longevity will keep growing, and it will face a situation where most of the population is composed of older people who will not be able to do all the heavy work needed. Minsky has delivered this argument in an interview with Kurzweil, who stands for the same idea.[48] But not everyone has this missionary purpose. The futurists certainly have.[49]

Lighthill missed the point that there were critical people in the front of Artificial Intelligentsia, starting with McCarthy. In his review of Lighthill's Report, he argues that the major problem, "the combinatorial explosion," was recognized since the beginning of the discipline and heuristics in its usual sense was the device for reducing the explosion. But he criticizes that "regrettably, some people were briefly over-optimistic about what general-purpose heuristics for theorem proving could do in problem-solving,"[50] and he also says that "it would be a great relief to the rest of the workers in AI if the inventors of new general formalisms would express their hopes in a more guarded form than sometimes has been the case."[51] Finally, he admits that "artificial intelligence has so far been only moderately successful."[52]

Since we are talking about people, their choices, and disputes, it is worth emphasizing a point raised by Pamela McCorduck, the historian of the discipline, from her interviews with the "Artificial Intelligentsia" about the report. She says the report was done in a hurry, and one reason for that, as the rumors of the time, say, it was due to "personal vendetta." McCorduck argues that "since the report coincided with, and undoubtedly exacerbated, the dissolution and reorganization of the Edinburgh University AI laboratory, which pioneered the research in AI in the United Kingdom, the rumors seemed right."[53] I would say this is a challenging position to be held since Sir Lighthill's arguments are sound, and some futuristic positioning could not be stopped. For example, during the Lighthill debate, Michie quoted the following report submitted on November 15, 1876, to the president of the U.S. Telegraph Company as a support for the inevitable future achievement of AI's goal:

> Mr. Hubbard's fanciful predictions while the sound very rosy are based upon wild-eyed imagination and a lack of understanding of the technical facts of the situation and a posture of ignoring the obvious technical limitations of this device, which is hardly more than a toy, a laboratory curiosity. Mr. AG Bell, the inventor, is a teacher of the hard of hearing, and this telephone may be of some value for his work, but it has too many shortcomings to be seriously considered as a mean of communication.[54]

Striking as it was on Lighthill in the debate, it was shortly after discredited by Professor Christopher Strachey (1916–1975) from the Oxford University, who was on the audience and argued against the "anecdotal method of supporting work in a very difficult field by quoting a totally irrelevant and incorrect prophecy about Bell's telephone a hundred years ago."[55] Lighthill appealed for the clarification role of the scientific community that "has a heavy responsibility to put forward its carefully considered view of the facts to avoid the public being seriously misled." In his view, the risk of AI was that it would mislead the public, "computers had been oversold," and the audience was very willing to have robots, and they were by no means against this idea. Saying so, his perception of the risk of AI as the technology was far away from the counterfactual scenario represented by Kubrick and Clarke's narrative, where HAL 9000 kills the spaceship crew because they were a threat for achieving the goal of the mission. Moreover, Lighthill was pretty much confident with his ideas; after all, for him, human beings were unique, as well as their brains. As we will see, the uniqueness of human capabilities is undoubtedly a tenet that group many of the oppositions to AI.

The aftermath of Lighthill's Report was that it provoked a massive loss of confidence in AI by the academic establishment in the United Kingdom, including the funding body. The pioneer country in the discipline was not the United Kingdom, but the United States, where the whole enterprise was being attacked as well. The philosopher Hubert Dreyfus (1929–2017) famous positioned against the goal of AI in his book *What Computers Can't Do*.[56] However, the main strike on The Artificial Intelligentsia budget came from the Vietnam War. At the end of 1969, the American Congress passed the "Mansfield Amendment," which enforced that research funded by the Defense Department should be strictly related to military functions or operations. By 1974, funding for AI projects became everywhere hard to find.

This situation persisted for almost a decade. This time, the favorable funding in the United Kingdom followed a good report known as "A Programme for Advanced Information Technology" or only "The Alvey Report." The next move of the discipline reinforced the issues raised about the term "artificial intelligence."

In 1973, the United Kingdom chose not to try to lead the world in AI, because the hype was fading away. Furthermore, the name became stigmatized not only in the United Kingdom but around the world. The magnetic power of the term was neutralized for an extended period. However, the manifold enterprises regarding AI persisted. As we know, they have been being attracted again under the name AI back at a point where almost forty-five years in the future, when a new hype is at the peak, the position the UK government has changed entirely.

In the United States, Hans Moravec, the prominent AI expert and futurist, described well how "things had gotten a little bit weird" by the early 1970s, Hans Moravec pointed to the role promising played in bringing about the funding cuts:

> Many researchers were caught up in a web of increasing exaggeration. Their initial promises to D/ARPA had been much too optimistic. Of course, what they delivered stopped considerably short of that. *But they felt they couldn't in their next proposal promise less than in the first one, so they promised more.*[57]

Moravec also added that ARPA, having been "duped," decided that "'some of these people were going to be taught a lesson,' were having their two-million-dollar-a-year contracts cut to almost nothing!"[58] AI was an investment risk, not a promising technology anymore. The credibility was lost and its ability to make predictions. Finally, the bubble was pierced, and the "AI winter" came.

NOTES

1. Manfred Clynes and Nathan Kline, "Cyborgs and Space," no. Astronautics (September 1960), 26.
2. Louis Fein, "Computer-Oriented Peace Research," *AFIPS'63* (Fall 1963), 631–639.
3. Louis Fein, "The Artificial Intelligentsia," *IEEE Spectrum* (February 1964), 74.
4. Fein, "Artificial Intelligentsia," 86.
5. Fein, "Artificial Intelligentsia," 86.
6. Fein, "Artificial Intelligentsia," 86.
7. Hubert Dreyfus, *Alchemy and Artificial Intelligence* (Santa Monica, CA: The RAND Corporation, December 1965), http://www.rand.org/pubs/papers/2006/P3244.pdf.
8. Fein, "Artificial Intelligentsia," 74.
9. David Stork, "Scientist on the Set: An Interview with Marvin Minsky," in *HAL's Legacy: 2001's Computer as Dream and Reality*, ed. David Stork (Cambridge: The MIT Press, 1997), 28.
10. The Computer Pioneer Award was established in 1981 by the IEEE Computer Society to recognize those people whose efforts resulted in the creation and development of the computer industry.
11. David Kuck, "Could We Build HAL? Supercomputer Design," in *HAL's Legacy: 2001's Computer as Dream and Reality*, ed. David Stork (Cambridge: The MIT Press, 1997), 50.
12. Stork, "Scientist on the Set," xix.
13. The ranking's criteria are: (1) if the movie has a narrative format with 60 minutes minimum duration, (2) if it is in English, (3) critical recognition, (4) major award winner, (5) popularity over time (including figures for box office adjusted for inflation, television broadcasts and syndication, and home video sales and rentals), (6) historical significance (technical innovation, visionary narrative devices or other ground-breaking achievements), and (7) the cultural impact.
14. Stork, "Scientist on the Set," 28.
15. The actual script evolved from a Clarke's short story Kubrick was interested in, "Sentinel of Eternity" (1951; aka "The Sentinel").
16. Stork, "Scientist on the Set."
17. Michael Mateas, "Reading HAL: Representation and Artificial Intelligence," in *Stanley Kubrick's 2001: A Space Odyssey. New Essays*, ed. Robert Kolker (Oxford: Oxford University Press, 2006), epub.
18. Arthur C. Clarke, *2001: A Space Odyssey* (New York: New American Library, 1968), chap. 16.
19. Clarke, *A Space Odyssey*, chap. 24.
20. Erving Goffman, *The Presentation of Self in Everyday Life* (New York: Anchor Books, 1959).
21. Roger Caras (1928–2001) was a publicist, VP of Stanley Kubrick's two companies, Hawk Films and Polaris Productions.

22. Michael Benson, *Space Odyssey: Stanley Kubrick, Arthur C. Clarke, and the Making of a Masterpiece* (New York: Simon & Schuster, 2019), epub.

23. One of the remarkable exceptions was Dendral, an early example of an expert system that we will explore in the next chapter.

24. Frank Rosenblatt, "The Perceptron: A Probabilistic Model for Information Storage and Organization in the Brain," *Psychological Review* 65 (1958), 386ff.

25. Nilsson, *The Quest for Artificial Intelligence,* 91.

26. Frank Rosenblatt, *Principles of Neurodynamics* (Washington, DC: Spartan Books, 1962).

27. Marvin Denicoff said, "At that time [in the 1960s], the Office of Naval Research had funds at the level of $40K or $50K. ARPA was able to fund hundreds of thousands, or even millions. Rosenblatt never attracted that kind of money, because he wasn't offering a large pay-off. By pay-off I mean not in the scientific sense, but in the application sense, world problem-solving. Again, his work was much more, I would say, traditional science. The Office of Naval Research never gave him the kind of money that he really required, and he was not successful in getting the money from the Science Foundation or from ARPA" (Olazaran, 1996, p. 638).

28. Nilsson, *The Quest for Artificial Intelligence,* 162.

29. Roger Schank and Robert Abelson, *Scripts, Plans, Goals and Understanding: An Inquiry into Human Knowledge Structures* (New Jersey: Erlbaum, 1977).

30. Nilsson, *The Quest for Artificial Intelligence,* 148.

31. James Lighthill, *Artificial Intelligence*, 2.

32. Lighthill, *Artificial Intelligence*, 8.

33. Lighthill, *Artificial Intelligence*, 13.

34. Hand-eye work was also being pursued at Hitachi's Central Research Laboratory in Tokyo during this period.

35. Lighthill, *Artificial Intelligence*, 12.

36. Lighthill, *Artificial Intelligence*, 13.

37. Lighthill, *Artificial Intelligence*, 12.

38. McCorduck, *Machines Who Think*, 118.

39. John McCarthy, John McCarthy (1927–2011): Artificial Intelligence, interview by J Mishlove, 1989, https://thinkingallowed.com/2jmccarthy.html.

40. BBC, "The General Purpose Robot is a Mirage," *Controversy* (London, 1973).

41. Brian Bloomfield, *The Question of Artificial Intelligence: Philosophical and Sociological Perspectives* (London: Routledge Kegan & Paul, 1988).

42. John McCarthy, "Review of the Question of Artificial Intelligence Edited by Brian Bloomfield" (History of Computing, Annals of the History of Computing, 1989), 6.

43. McCarthy, "What is Artificial Intelligence," 2.

44. John McCarthy, "Review of 'Artificial Intelligence: A General Survey' by Professor Sir James Lighthill," *Artificial Intelligence* 5 (1974), 2.

45. McCarthy, "What is Artificial Intelligence."

46. John McCarthy, John McCarthy Oral History, interview by Nils Nilsson, September 12, 2007, https://www.computerhistory.org/collections/catalog/102658149.

47. Lighthill, *Artificial Intelligence*, 12.

48. Marvin Minsky, Kurzweil Interviews Minsky: Is Singularity Near?, interview by Ray Kurzweil, 2010, https://www.kurzweilai.net/kurzweil-interviews-minsky-is-singularity-near.

49. As we will see, many futurists will embrace the idea of "singularity" reshaped and propagated by Kurzweil.

50. McCarthy, "Review of Artificial Intelligence," 11.

51. McCarthy, "Review of Artificial Intelligence," 12.

52. McCarthy, "Review of Artificial Intelligence," 12.

53. McCorduck, *Machines Who Think*, 244.

54. BBC, "The General Purpose Robot."

55. BBC, "The General Purpose Robot."

56. Hubert Dreyfus, *What Computers Can't Do: A Critique of Artificial Reason* (Cambridge: MIT Press, 1972).

57. Crevier, *AI,* 115.

58. Crevier, *AI*, 117.

Chapter 4

We Thought We Could Do It

Human-Machine Symbiosis, Heavy Investments, and the Raising of Futurists in the 1980s/1990s

In the previous chapter, we saw that the Sputnik crisis had boosted a wave of fears that led to major social impacts based on the risk amplification dynamics, as we have presented in chapter 1. In the following period, from the 1980s to the early 1990s, there was another political imaginary pulling force in the United States and Europe represented as a future risk, the "Japanese threat." It regards the *Fifth Generation Computer Systems*, a Japanese national plan to develope intelligent computers. Contrasting with the Sputnik crisis, in this case, there was no physical triggering event in Japan because it did not achieve the goal of producing this sort of computer. As we have already learned, future risks may not have any physical correspondent event in artificial intelligence (AI) history. The conceived risk scenarios are enough to put forward the process of amplification of risk representations and their associated secondary social impacts. Once again, experts are in charge of the representation of future risks and communicating a legit version for the media. They came to the public not merely for educating the laypeople, but especially for communicating the future risk and the possible course of actions that should be taken. In this case, the Japanese threat fostered the American, British, and European AI experts to take action and try to get funding for further research.

During this period, the term "artificial intelligence" got some variations. A peculiar trait of this period is the encompassing of the word and the idea related to "knowledge" as the new paradigm for the discipline. In what follows, instead of being advertised as AI, there were some other names that appeared for representing the same idea, for example, *knowledge information processing systems* or *knowledge-based systems*. Experts systems were often mixed up with AI in general—similarly to what happened today with the term "machine learning" and "artificial intelligence." Of course, the term "robots" remained existing throughout this period.

According to Edward Feigenbaum and Pamela McCorduck's survey about the worldwide AI activity, in the 1980s, there were sixty-three organizations dedicated to it in a wide variety of application areas. According to their classification, the most representative areas were still Robotics (nineteen organizations), General AI (seventeen organizations), Expert Systems (seventeen organizations), and Fifth Generation Computer (five Japanese only organizations). This survey also showed how AI was still chiefly an American specialty. Of the sixty-three total organizations, fifty-three were in the United States (around 84 percent), six of them were in Japan, three were in the United Kingdom, and one was in France.[1] In this context, Feigenbaum was probably the leading public character advocating for research funding in the United States. This renewed research would bring not applications for toy problems anymore, but most expert systems designed to help actual work. According to Nilsson's history of the field,[2] the successful applications encouraged specialization into subdisciplines. Concerning to real threat the general public felt from AI, it was still far away, belonging to fictional universes. At the end of this period, we will see the rise of the internet, certainly the second major technological event in our context after the creation of the computer.

The origin of the internet traces back to the previous period we analyzed, usually considered the ARPANET project as the first one (1969). In 1960, Joseph Licklider wrote a paper that represented his vision on what would be ARPANET and eventually, the internet. In 1962, Licklider was appointed to be the head of ARPA's information processing techniques office; and he was the one in charge of approving the funding for symbolic AI research. His 1960s book may help us to understand why he was so optimistic about this research and how it was related to his vision of the future internet. The title of the paper was *Man-Computer Symbiosis*, which stood for the idea that there should be a partnership of very close coupling between the human and the electronic members. Such a partnership would *mechanically extend man*, and computers may "outdo the human brain in most of the functions we now consider exclusively within its province."[3] He backed his claim with some early achievements of the research in AI we presented before, such as problem-solving and pattern recognition programs. It means that computers had already been "capable of rivaling human intellectual performance in a restricted area." He also conceived a (possible) scenario of "dominance in the distant future of cerebration to machines alone," but he did not discuss that in his text. However, a requisite for this kind of future scenario that he depicted was buttressed on the development of something like the internet,

> It seems reasonable to envision, for a time 10 or 15 years hence, a "thinking center" that will incorporate the functions of present-day libraries together with anticipated advances in information storage and retrieval and the symbiotic functions suggested earlier in this paper. The picture readily enlarges itself into a network of such centers, connected to one another by wide-band communication lines and to individual users by leased-wire services. In such a system, the speed of the computers would be balanced, and the cost of the gigantic memories and the sophisticated programs would be divided by the number of users.[4]

During the 1980s, the networks were merged. For instance, the National Science Foundation's network CSNET (1980) had been connected to the ARPANET supported by a "Internet Protocol." Besides, before the establishment of the NSFNET in 1985, this merge laid down the backbone of the internet, which became the host of all the sites forming the World Wide Web.

These changes replaced the "cosmic breakout" motif with different breakthroughs based on a new and more exciting life as pioneers in the *cyberspace*. Not only an expert as Licklider foretold this invention, but also foretold the science-fiction champion, Arthur C. Clarke, the *2001 Odyssey's* writer. In the course of an interview for the Australian Broadcast Corporation (ABC) in 1974, Clarke said that in the year 2001 there would be home computers of the size of a console that would be capable of communicating to other computers and retrieving everyday information needs such as bank statements, theater reservations, and so on. In contrast to the expert's forecast, Clarke conceived the consequent societal impacts of this technological *novum*. When the interviewer asked him about social consequences, he exposed the scenario where home computers connected to others would enrich social life, enabling someone to live anywhere in the world.

As a consequence, it would change the way we do business. The technology would also keep expending and becoming pervasive in social life. Once again (social science) fiction seemed to be ready to conceive the further consequences of this kind of achievement. However, there was no paradigmatic cultural representation of the development of the technology and its derived risks in this period, as we saw with Kubrick's *2001*. The representation of risks at this moment conceived less possible scenarios in which a long chain of events (breakthroughs) are taken for granted. The aim was exploring not only the social consequences but also, for the first time, the psychological consequences of the progress of AI in the form of robots or pure software. This was the main topic for the cyberpunk movement that arose in the 1980s.

Within this context, the writer William Gibson coined the concept of "cyberspace" in his book *Neuromancer* (1984). The idea of virtual space was not something new in fiction. For instance, Daniel F. Galouye's *Counterfeit World* (1964), Chris Boyce's *Catchworld* (1975), and Vernor Vinge's *True*

Names (1981) had already depicted the idea. Still, in the *Neuromancer*, the consequences of this scenario of emerging technology and its significant results were unfolded. The concept of "the matrix" also came forth. It had been used before in Douglas R. Mason's *Matrix* (1970) and appeared again in no small degree in Gibson's *Neuromancer*. We are familiarized with this idea today because of the movie, the Wachowskis brothers' *The Matrix* (1999). The cyberpunk way of conceiving scenarios also appeared in animations, especially in Japan. Probably, anywhere else in the world, a population had been so impacted by the fast pace of technological innovation. Katsuhiro Otomo's *Akira* (1988) and Mamoru Oshii's *Ghost in the Shell* (1995) represented that even though they have started much earlier in the *manga* medium—for example, Akira (1982).

The cyberpunk fiction popularized how AI might bring profound changes in human lives, such as a new kind of afterlife as a consequence of starting to think about a virtual existence. The feeling raised by how we represent the risks from AI became more nuanced, it moved from an acceptance or rejection position, and economic and social impacts, to the excitement, anxiety, despair, depression, and consequently subjective side effects. Now AI hits human identity. This movement has started during this period and keeps existing until the present day—contemporary movies such as Jonze's *Her* (2013) and Alex Garland's *Ex Machina* (2014) explore it in a more realistic and subtle way.

Even though the novelty of this category of risks, the potential subjective threats brought by AI appeared since the beginning of the discipline. One of the inventions of the previous period, the first chatbot ELIZA, designed by Joseph Weizenbaum (1923–2008), showed how people could be easily deceived by AI in a superficial communication. The chatbot simulated a conversation by using a simple pattern matching and substitution methodology, and still, it managed (according to Weizenbaum) to deceive some human users. Contemporary, one of the founding fathers of virtual reality, Jaron Lanier, has been advocating against the illusion brought by some AI applications.

The cyberspaces become the inner space of a much vaster mind. We can imagine the human relationship with AI in these places, and it must exert some psychological or emotional impact on human's psyche. The dormant *cyber* vision of the world will flourish not only in the science-fiction literature but also in the futurists' interventions that may also come from an expertise-related background—for instance, the Austrian computer scientist (and actual Carnegie Mellon professor) Hans Moravec and the American computer scientist Ray Kurzweil. These futurists began brewing the risks from super-intelligence in this period, even though the initial kick was made by the English mathematician I. J. Good (1965), who has worked with Alan

Turing in cryptography projects. The futurists will also hold as a classic the science-fiction novel *True Names* by Vernor Vinge. Both cyberpunk ideas and the futurists making their interventions in nonfictional contexts will survive up to the present time. Old futurists (and experts), such as Minsky, will reappear and will also show short memory concerning the field.

Stork: Back to the mid-sixties, wouldn't you agree that the field was quite optimistic? After all, in a Life magazine article, you were quoted as saying "In from three to eight years we will have a machine with general intelligence of an average human being. I mean a machine that will be able to read Shakespeare, grease a car, play office politics, tell a joke, have a fight. In a few months it will be at genius level and a few months after that its powers will be incalculable.

Minsky: Oh, that Life quote was made up. You can tell it's a joke. Herbert Simon said in 1958 that a program would be chess champion in ten years, and, as we know, the IBM group has done extremely well, but even now [Deep Blue] is not the undisputed champion.[5]

This period brewed the current visions of the future of AI. It differs from the previous period in terms of the reach enabled by the internet to have access to such scenarios. It also enabled experts to convey those ideas without creating a fictional world, and without the barriers of academic peer reviews. In what follows, these future scenarios become a sort of entertainment.

CYBERPUNK CULTURE

Perhaps, we may say that the "cyberpunk" regards future scenarios in which we have a dramatic situation that high technology contributed to a world of social inequality. The idea that these scenarios are "dramatic" means that they have been explored almost to its limits. In this sense, we always have some kind of dystopian scenario, where the social order is in collapse.

Ridley Scott's *Blade Runner* (1982) became the primary reference of the cyberpunk narratives and aesthetics. As regard the visual imagery, we have references to Lang's *Metropolis* but chiefly from the French comic strips, especially Jean Giraud (aka. Moebius). It differs in many aspects of what Kubrick sought in *2001* in the sense of conceiving a counterfactual world closest as possible to our world. In *Blade Runner*, many narrative links in the chain had already happened. So, it is impossible to access how we reached that situation. The androids, called the *Replicants* in the movie, are not the cognitive *novum*, rather the androids' emergence of emotions. Besides the imagery of the dystopian world, the narrative gravitates around the emotions and the question about human identity.

The most important device is the Voight-Kampff polygraph-like machine, a fictional interrogation tool, used by blade runners to determine whether or not an individual is a replicant. It measures bodily functions such as respiration, blush response, heart rate, and eye movement in response to questions dealing with empathy. In a sense, questions are always about imaginary worlds with an emotional trait. The first string of questions of the Voight-Kampff in the movie portrayed the following scenario: "You are in the desert walking along the sand and all of the sudden you look down, and you see a tortoise, it's crawling towards you. The tortoise lays on his back. Its belly is baking in the hot sun beating its legs, trying to turn itself over, but it can't. Not without your help. But you are not helping." The replicant reacted literally too many aspects of this scenario. For example, he started to question "what desert?" and the interviewer says, "It makes no difference the desert is completely hypothetical." The replicant replied, "How come I'd be there?" and the interviewer replied that "maybe you're fed up, maybe you wanted to be by yourself, who knows!?" It is clear from this dialogue that the replicant could not deal with these hypothetical situations. The replicant also questions, "What do you mean I am not helping?" in a rush of anger.

The line between humans and robots is drawn by empathy and imagination. However, throughout the movie, Rachael, a more advanced replicant, need more than a hundred questions to determine whether or not she was a robot. The boundary between humans and robots was pushed out farther. There is no limit for that. For instance, there is also an unanswered question about whether or not Deckard, the protagonist, is a robot. However, Blade Runner's points are not referred by experts as a goal. The same may be applied to all AI representation in this cyberpunk genre. Gibson's *Neuromancer* published two years after *Blade Runner*'s release represented a gifted hacker with a belonging feeling for cyberspace instead of the real world. The story depicts the development of AIs (Wintermute and Neuromancer) but limited by a hypothetical Turing Law Code, which bans the construction of this level of AIs.

An important current influence on different domains is the idea of technological singularity devised by Irving Good in the 1960s and popularized by Vernor Vinge. In contrast with the dystopian cyberpunk's scenarios, Vinge is an optimist. For him, progress with human/computer interfaces is effectively Intelligence Amplification[6] (Marcus et al., 1992). Vinge's *True Names* (1981) had already depicted the cyberspace, virtual reality, but in a more realistic setting. It explored less the consequences of a network of computer and AIs that may inhabit it for human society and psyche than the extrapolation of the practices of real young hackers subculture. Important experts such as Marvin Minsky and Daniel Hillis wrote introductory essays to this book. Vinge used the expression "technological singularity" in his novel *Marooned in Real-Time* (1986). The term singularity is originally a mathematical term to describe critical points at which functions become indefinable, usually by tending to infinity. The American science-fiction writer Mildred Downey Broxon imported it in his stories called *Singularity* (1978). This idea means that the evolution of technology will continue to accelerate to a point that will bring an abrupt and dramatic breakthrough. Damien Broderick, an Australian science-fiction writer, also developed a similar notion of technological singularity in novels at the end of this period, *The Spike: Accelerating into the Unimaginable Future* (1997) and *The Last Mortal Generation: How Science Will Alter Our Lives in the 21st Century* (1999). This idea will be adopted by futurist movements, such as transhumanism and extropian movements in the 1990s.

HEAVY INVESTMENTS ON AI

Japanese economy boomed throughout the 1950s, 1960s and the 1970s. It was called the Japanese miracle. From 1953 to 1973, the Japanese economy multiplied by more than five times. In 1982, Japan's *Ministry of International Trade and Industry* launched a national plan called the *Fifth Generation Computer Systems*. They aimed at improving computer technology to leap Western computer expertise by creating an entirely new computer technology capable of performing inferences from knowledge bases as well as improving the natural language. Developing such technology meant that developing many parallel processors using ultra-large-scale integration. The project was organized around basic research, and for such undertaking, the Ministry created a special institution, the *Institute for New Generation Computer Technology* (ICOT), a consortium backed by eight Japanese firms, Fujitsu, Hitachi, NEC, Mitsubishi, Matsushita, Oki, Sharp, and Toshiba as well as national laboratories, the Nippon Telephone and Telegraph's Musashino Laboratories, and the Ministry of International Trade's Electrotechnical Laboratory.

The scientists of this institute chose PROLOG as the primary programming language for the project instead of LISP, widely used in the United States during this period. The institute managed to build some parallel inference machines. According to researchers of the project, "the research and development principle throughout the project has been to adopt logic as the theoretical backbone of knowledge information processing and parallel processing as the key technology for obtaining high performance"[7] (Chikayama and Yokohara, 1993, p. 337). However, as Nilsson said, "Taking full advantage of the benefits of parallel processing proved difficult except for special problems susceptible to that style of computation."[8]

The development was related to expert systems, which are computer programs that perform specialized, professional tasks. They rely on large bodies of knowledge; therefore, they are also known as knowledge-based systems. They need knowledge engineering. Knowledge is the key factor in the performance of an expert system. It can be either a fact of the domain or heuristic knowledge, which is the knowledge of good practice and judgment in a field.[9] The system uses a knowledge base. In a medical context, it would be analogous to what one learned during the medical education and training (medical school, residency, internship, practice as well as articles in journals, etc.). Since this kind of knowledge is implicit for the expert, someone else has to mine it. The miners are the *knowledge engineers,* AI people who know how to represent knowledge digitally. Expert systems have three problems: knowledge representation, knowledge utilization, and knowledge acquisition.[10]

Japan's Challenge transmuted the launch of the Japanese project as a threat akin to the launch of *Sputnik* decades earlier. In response, a group of American companies formed the Microelectronics and Computer Technology Corporation (1983), a for-profit consortium of thirteen corporations in Austin, Texas. And the Defense Department began a massive long-term program to develop intelligent systems, including tanks that could navigate on their own, the Strategic Computing Initiative. It was a billion-dollar, decade-long DARPA-funded computing project. Engendering international projects such as MCC, Alvey, and European Strategic Program of Research in Information Technology, was a token of success of the Japanese enterprise.[11]

Feigenbaum and McCorduck book *The Fifth Generation* expressed "They're going to give the world the next generation—Fifth Generation—of computers, and those machines are going to be intelligent"[12] whereas other countries would be exporting food, commodities, and other manufactured goods. "When no corporation or foundation chose to take AI seriously or could afford to, the ARPA of the Department of Defense supported it through two decades of absolutely vital but highly risky research . . . the Pentagon is often perceived as the national villain, especially by intellectuals."[13] "We believe that Americans should mount a large-scale concentrated project of our own; that not only is it in the national interest to do so, but it is essential to the national defense."[14] Feigenbaum calls to action eventually mentions Pearl Harbor, as an American failure of intelligence. Six alternatives for America: (1) maintain the status quo; (2) form industrial consortiums; (3) enter a joint venture with the Japanese; (4) software only production; (5) form a national laboratory; (6) "prepare to become the first great agrarian postindustrial society."[15] The national laboratory would be the best choice, according to them.

Explicitly motivated by Japan's national plan, the British government set up a committee to advise on the scope for a collaborative research program in information technology (IT) and to make recommendations. They saw both the Japanese application and United States' response to it as a threat to the

UK industry, so it was recommended a five-year program and £350 million in "perhaps the first example of collaboration between UK academia, industry, and government on a major IT research project."[16] The four major technical areas identified for support were "Software Engineering, Man-Machine Interfaces (MMI), Intelligent Knowledge-Based Systems (IKBS), and Very Large-Scale Integration (VLSI)." What was the IKBS? Kenneth Owen, an editorial consultant to the Alvey Directorate from 1983 to 1988 tells the following story,

In the summer of that year, Britain's artificial intelligentsia struck back. "A bunch of us had felt quite a time that we really ought to try to do something about rehabilitating artificial intelligence in the UK after Lighthill," says John Taylor, at that time Head of the Command Systems Division of the Admiralty Surface Weapons Establishment and also Chairman of the Computing and Communications Subcommittee of the Science and Engineering Research Council (SERC) IEC. The aim was to set up a SERC "specially promoted program" in AI, along the lines of the Distributed Computing Systems project. At an early meeting in London, a group including Donald Michie, a pioneer of AI research at Edinburgh University, and Robert Kowalski of Imperial College discussed strategy. "We were sitting around a table wondering what to call this new area," says John Taylor. "Should we call it artificial intelligence? We didn't want to call it artificial intelligence because of all the Lighthill connotations, and we didn't want to call it expert systems, and we came up with this awful phrase 'intelligent knowledge-based systems' or IKBS."[17]

Most of the funding under the Alvey program came to an end around 1988. According to Robert Kowalski, the program supported thirteen research grants for our group, involving a total expenditure of £1.5 million. At its peak in 1987, it had approximately fifty researchers.[18] A similar European project named as ESPRIT, set up to support research in three major categories: microelectronics, information processing systems, and applications. The AI-related projects developed various knowledge-based systems, logic programming environments, natural language parsing systems, and knowledge acquisition and machine-learning systems. After the end of Alvey funding, many British researchers collaborated to ESPRIT.

The aftermath of the project is that it aimed at the wrong target. The industry moved to the direction of developing *graphical user interfaces*, enabling laypeople human–computer interaction. It lowered the requirements for natural language and speech understanding systems. After they had spent ten years and US$380 million in 1990 rate of exchange (US$747 million in 2019), the project failed.

The Japanese researchers were, nonetheless, conscious and cautious regarding their plan, which was amplified during the spread of the Japanophobia in the United States. Feigenbaum and McCorduck's book is a good example of such phobia. We can find some stereotypical statements there, "Although Japanese managers have taken on a samurai-like glamour in the past few years, thanks to enviable worldwide successes, they are in general conservative and risk aversive. Yet here they are, charged with managing a very ambitious, high-risk project based on the technology they hardly understand."[19] They also partially reproduce some stereotypes of the time, such as the Japanese Economy is a perfect copycat—look how familiar sound those arguments for today's ears concerning China.[20] However, they were trying to alert that this belief was not corresponding to the truth anymore.

Ten years after the beginning of the project, in a joined article called *Launching the New Era,*[21] Kazuhiro Fuchi, the director of the research center of ICOT, declared,

> When this project started, an exaggerated image of the project was engendered, which seems to persist even now. For example, some people believed we were trying, in this project, to solve in a mere 10 years some of the most difficult problems in the field of artificial intelligence(AI), or to create a machine translation system equipped with the same capabilities as humans. In those days, we had to face criticism, based on that false image, that it was a reckless project trying to tackle impossible goals. Now we see criticism, from inside and outside the country, that the project has failed because it has been unable to realize those grand goals.[22]

Robert Kowalski, a researcher for the Imperial College at the end of the project's time, added that "Not only were the original FGCS targets for AI exceedingly ambitious, but they were considerably *exaggerated* by some commentators, most notably perhaps by Feigenbaum and McCorduck in their book *The Fifth Generation.*"[23]

Besides the assessment of the past ten years of the project and the misperceptions expressed by Western scientists, we should stress which believes were under the idea of *New Era.* The end of this project was seen as the Japanese contribution "toward the future of the human race," Kazuhiro also added in sequence, "we believe we are responsible for leaving our research accomplishments as a legacy to future generations. . . . The advancement of computers and information processing technologies is closely related to the future of human society. Social thought, ideologies, and social systems that fail to recognize its significance will perish, as we have seen in recent world history."[24]

THE AGE OF FUTURISTS

A futurist is someone who communicates to the public his or her counterfactual scenarios of the future and their consequences in a nonfictional context.

The frontman of this movement will become Ray Kurzweil at the end of the 1990s. A great number of other experts and futurists (or both) gathered around him. Before him and in parallel to him, many other experts came to the public for communicating future scenarios related to the achievement of AI.

One of the most important communications about future scenarios for a wider public is Hans Moravec's article for the *Wired* magazine entitled *Why the future does not need us?* (1998). Hans Moravec is known for his work on robotics, AI, and writings on the impact of technology at the Carnegie Mellon University. He had been at the forefront of robotics research for some decades. He was also John McCarthy's PhD student at the Stanford University from 1971 to 1980.

In an informal report named *Intelligent Machines: How to get there from here and What to do afterward* (1977), he diagnosed why AI was not matched yet, "it is simply that the machines with which we are working are still a hundred thousand to a million times too slow to match the performance of human nervous system."[25] Hal 9000 is included in his acknowledgments as a source of inspiration. In the final section of the paper, named "The Future," he conceives future scenarios and the consequences from there. He thought we might have enough computing power in about ten years (1987) for the current price of a medium-large computer if the software kept the pace of development, "machines able to think as well as humans begin to appear in 10 years."[26] Society would have been fulfilled with intelligent machines with enormous advantages over humans in competitive situations. The successive generations of machines produced this way will be increasingly smarter. "Regardless of how benevolent these machines are made, homo sapiens will simply be outclassed." This point of the narrative is familiar to us. He continues with his future scenarios fulfilled with machine space journeys, colonies, and so on. There would exist a machine society.

He kept developing his future scenarios in a somewhat fatalist tone. So the risk of human beings superseded by machines must be understandable. The *Intelligent Machines'* report was reprinted as a book chapter with some changes, *The Endless Frontier and The Thinking Machine* (1982). In our analysis, we will consider Moravec as the first expert who will inaugurate the futurist view. He explicitly positions himself as such. Unlike Marvin Minsky who conveyed his future scenarios now and then, so he remained an expert above all, Moravec kept both positions: the expert and the futurist. He started conveying the future scenarios derived from imaginable consequences of technology for a broader audience in a book called *Mind Children* (1988). He updated his previous beliefs of the period when robots would show human-like abilities, within fifty years or so. A year before, he assumed his public positions in a paper called *Harvard Doesn't Publish Science Fiction*,

> *Mea culpa*: reality and fantasy blend in my mind. My major source of income is as a scientist, and, fortunately for all involved, on my very best behavior I can pass as such. What's more, it's fun. Working out problems slowly, carefully and rigorously gives insight, a sense of solidity, and of a job well done. Often it sheds light in unexpected directions, and opens new doors. But it has limitations. To do it, you must already understand the problem in a detailed way. But many interesting questions are too nebulous and slippery for such intimacy. My burden is that often I'm insensitive to the difference.[27]

Mind Children started the propagation of his ideas outside Academia. The book was reviewed by the *Washington Post, New York Times*, and *The New Yorker Magazine*. The *New York Times'* review presented witty comments that reveled much of our argument. M. Mitchell Waldrop, a reporter for *Science* magazine and author of the book *Man-Made Minds:* a study of artificial intelligence wrote the review. We will see some of his arguments shortly.

For the next decade, he worked on the idea of the Universal Robot, showing updated predictions, and kept communicating his ideas also in transhumanism venues. For example, he gave a talk called *The Age of Robots* (1994) for the *First Extropy Institute Conference on TransHumanist Thought*, invited by Max More, a British futurist. The publication was based on his text of the Universal Robot. In 1998, he was invited by Nick Bostrom for publishing a paper for the *Journal of Evolution and Technology* named *When Will Human Hardware Match the Human Brain?* In the same year, he published another book for a wider audience called *Robot: Mere Machine to Transcendent Mind* (1998), and in the next year, he wrote *Rise of the Robots*, in *Scientific American* (1999). *Robot* was reviewed by a number of important media venues such as *the New York Times, Washington Post*, *Wired*, and *Nature*.

> While many readers will doubtless find this prospect terrifying, Mr. Moravec glories in it. The robots he envisions in *Mind Children* will be nothing at all like the rigid, emotionless automatons we have come to expect from bad science fiction movies. They will be our companions, our helpmates, and our heirs. . . Well, all this is fun, not to mention mind-expanding. But then, so is *science fiction.* What makes Mr. Moravec's book different is that *he wants us to take it seriously*, as a realistic sketch of our not-very-distant future.[28]

The title of the Waldrop's review, *The Soul of the New Machines*, also anticipated the title of the famous Ray Kurzweil's book, *The Age of Spiritual Machines* (1999). Waldrop's review also criticized Moravec's book for not elaborating on the societal consequences of his vision of the future. In Asimov's parlance, it is a science fiction instead of social-science fiction. He does not "express the slightest doubt that what can be done will be done. . . . He says very little about the impact of humanlike robots on human employment—or, more importantly, on a humankind's sense of purpose and dignity. Apparently, we are simply to fade away, leaving the universe to our heirs and betters."[29] Interestingly, Waldrop brought into consideration a side effect of lack of purpose and dignity explored further in cyberpunk literature.

Nature, the scientific journal, wrote a single review for Moravec's book and Ray Kurzweil's *The Age of Spiritual Machines* (1998). John Casti, the reviewer, writes that the books "tells much the same tale" and "draws essentially the same conclusions." Casti closes his review by recommending "both books to any reader who wants a mind-expanding account of the rise of the age of intelligent machines" (Casti, 1999). However, McGinn went even further and did a joint review of three books: Moravec's, Kurzweil's, and Neil Gershenfeld's[30] *When Things Start to Think* (1998) under the witty name "*Hello, HAL,*" and a wittier opening,

> Has the invasion already begun? . . . Are machines, the products of human engineering intelligence, poised to take over the world—or is this an irrational fear, the latest spasm of the Luddite spirit? Finally, is the whole idea just a clever marketing ploy for the investment-hungry artificial intelligence industry?[31]

The joint impression presented in *Nature* and *New York Times* tells that there was some well-defined representation of the future as a consequence of the development of machine intelligence. Experts and futurists are the best future's heralds. They conceive scenarios but in a nonfictional context. These positions about the future a progressively propagated.

An intriguing question is why Kurzweil's book was much more popular provided it told a similar story when we compare to Moravec's book. Furthermore, Moravec had communicated these ideas much earlier than Kurzweil's first book. We can find a hint for that in the same McGinn's review for the *New York Times.* Kurzweil would have been a better communicator, and he was committed to develop a social-science thinking style of thinking. According to McGinn,

> [Kurzweil] develops the same kinds of speculations as Moravec, but with more of an *emphasis on the meaning of such innovations for human life* . . . he envisages the eventual triumph of the virtual over the real. His book ranges widely over such juicy topics. . . . This is a book for computer enthusiasts, science fiction writers in search of cutting-edge themes and anyone who wonders where human technology is going next.[32]

Kurzweil was Marvin Minsky's student at MIT in the late 1970s. He did not become a scholar but a businessman. His first book was *The Age of Intelligent Machines* (1990), which has also texts from leading figures related to AI of that time, for instance, Allan Newell, Daniel Dennett, Douglas Hofstadter, Edward Feigenbaum, Margaret Boden, Marvin Minsky, Seymour Papert, and Sherry Turkle. As a businessman, Kurzweil conceives scenarios as his business products. His next book speaks for itself, *The 10 percent Solution of a Healthy Life: How to reduce fat in your diet and eliminate virtually all risk of heart diseases and cancer* (1993). *The Age of Spiritual Machines* managed to be sold, but at the beginning of the 2000s (which marks the end of the period we are considering in this chapter), he published a best-seller book called *The Singularity is Near*, which finally popularized the future scenario of the rise of AI and its consequences. It has a similar narrative to his previous book and to Moravec's book, in the sense that once singularity is reached, machine intelligence will supersede human intelligence and will radiate outward Earth.

While in Moravec's or Kurzweil's words these future scenarios are neutral or optimistic, the affective bias can change the content of an imagined scenario. For example, in April 2000, the computer engineer and co-founder of *Sun Microsystems*, Bill Joy, wrote an article to the *Wired* magazine called *Why the future does not need us* (2000) in which he depicted a risk representation of machine intelligence buttressed on the same examples provided by Moravec and Kurzweil. Joy's intervention is useful for us to understand the influence of some experts such as Kurzweil. He declared in the article that "I had always felt sentient robots were in the realm of science fiction. But now, from someone I respected, I was hearing a strong argument that they were a near-term possibility."[33] He got the same idea from another expert and futurist, Danny Hillis, a Minsky's student who worked with parallel computing. He shored up Kurzweil's view, telling that all the changes imagined would come gradually and we would be used to it.

While Joy was reading Kurzweil's book (1999), he was captured by a dystopian example of the dystopian vision of Theodore Kaczynski,[34] the Unabomber, described as unintended consequences, a well-known problem with the design and use of technology.

Joy imagined that the threats were already portrayed in cyberpunk stories. In an overgrowing complexity in society, we would need to rely on machine decisions increasingly, and at some point, machines will be willy nilly in control. There would be a so strong dependence to machines that turning machines off "would amount to suicide." A tiny elite would be in control of the big machines and consequently, they will hold humanity in their hands. Actually, "masses will be superfluous, a useless burden on the system." The difference now is that these chains of imagined scenarios are self-proclaimed as nonfiction. In fiction context, we have the suspension of the misbelief. As regard stating the truth, if we remember Lewis's argument on truth in fiction, what the authors write in a fictional world has to be considered as a truth in that world. But when futurists, more or less prepared, come to the public, such condition is ignored and the truth is believed to be in the actual world.

NOTES

1. Edward Feigenbaum and Pamela McCorduck, *The Fifth Generation* (Reading: Addison-Wesley, 1983).
2. Nils Nilsson, *The Quest for Artificial Intelligence.*
3. J. Licklider, "Man-Computer Symbiosis," *IRE Transactions on Human Factors in Electronics* HFE-1 (1960), 4–11.
4. Licklider, "Man-Computer Symbiosis."
5. Stork, "Scientist on the Set," 19.
6. Aaron Marcus, Donald Norman, Rudy Rucker, Bruce Sterling, and Vernor Vinge. "Sci-Fi at CHI: Cyberpunk Novelists Predict Future User Interfaces," in *Proceedings of the SIGCHI Conference on Human Factors in Computing Systems* (*CHI '92*). Association for Computing Machinery, New York, (1992) 435–437.
7. T. Chikayama and R. Kiyohara, "Parallel Inference System Research in the Japanese FGCS Project," in *Parallel Symbolic Computing: Languages, Systems, and Applications*, ed. R.H. Halstead and T. Ito, Vol. 748, Lecture Notes in Computer Science (Berlin: Springer, 1993), 337–351.
8. Nilsson, *The Quest for Artificial Intelligence*, 351.
9. Feigenbaum and McCorduck, *Fifth Generation.*
10. In the PBS TV show *The Computer Chronicles,* Feigenbaum expresses a core idea of the experts systems. "We believe [the expert systems] will not be used unless they can explain to the human user how they are reasoning. So we have these explanation programs as pieces of knowledge engineering in the system that allow the system to be a transparent box; we shun black boxes" (Feigenbaum, 1984). In this sense, it is opposed to the current neural network approach.
11. Kazuhiro Fuchi et al., "Launching the New Era," *Communications of the ACM* 36, no. 3 (1993), 49–100.
12. Feigenbaum and McCorduck, *Fifth Generation,* 12.
13. Feigenbaum and McCorduck, *Fifth Generation,* 215.
14. Feigenbaum and McCorduck, *Fifth Generation,* 216.
15. Feigenbaum and McCorduck, *Fifth Generation,* 226.
16. Brian Oakley and Kenneth Owen, Alvey: Britain's Strategic Computing Initiative (Cambridge: MIT Press, 1989), 13.
17. Oakley and Owen, Alvey: Britain's Strategic, 15.
18. Fuchi et al., "Launching the New Era."
19. Feigenbaum and McCorduck, *Fifth Generation,* 120.

20. In a very weird story, Feigenbaum and McCorduck reproduced what a Japanese executive told them, "You Westerners . . . are the hunters of the world. You go out and find things, hunt them down. We Japanese are the cultivators, the agrarians. . . . We don't innovate. We don't hunt for something new" (Feigenbaum and McCorduck, 1983, p.137).

21. Fuchi et al., "Launching the New Era."

22. Fuchi et al., "Launching the New Era," 49.

23. Fuchi et al., "Launching the New Era," 56.

24. Fuchi et al., "Launching the New Era," 54.

25. Hans Moravec, "Intelligent Machines: How to Get There from Here and What to Do Afterwards," September 1977, https://frc.ri.cmu.edu/~hpm/project.archive/general.articles/1977/smart.

26. Moravec, "Intelligent Machines."

27. Hans Moravec, "Harvard Doesn't Publish Science Fiction New Destinies, Volume III, Jim Baen, Ed., Baen Books, February 1988, Pp. 212–247," in *New Destinies*, ed. Jim Baen, vol. III (Baen Books, 1988), 212.

28. Mitchell Waldrop, "The Souls of the New Machines," *New York Times*, January 1, 1989, https://frc.ri.cmu.edu/~hpm/book88/reviews/890101.NYT.html.

29. Waldrop, "Souls of New Machines."

30. Gershenfeld was the director of the celebrated Media Lab at the MIT.

31. Colin McGinn, "Hello, HAL," *New York Times*, January 3, 1999, https://archive.nytimes.com/www.nytimes.com/books/99/01/03/reviews/990103.03mcginnt.html.

32. McGinn, "Hello, HAL."

33. Bill Joy, "Why the Future Does Not Need Us?" *Wired*, December 1999, 124–35.

34. A scholar who abandoned the academic career in 1969 to pursue a primitive lifestyle. Between 1978 and 1995, he killed three people and injured twenty-three others in an attempt to start a revolution by conducting a nationwide bombing campaign targeting people involved with modern technology.

Chapter 5

Are We Almost There?

Cutting-Edge AI Developments, Social Media, and Politics in the Twenty-First Century

In the last two decades, the experts finally took the lead in conceiving the risk scenarios of artificial intelligence (AI) and communicated that to the media. The internet and social network services—SNS—(aka. social media) even enabled these experts to communicate directly to the public. As we have just seen, Moravec and Kurzweil pushed this movement that took traction during this last period. It happened because of a conjunction of factors chiefly related to the development of the internet and its tools.

In the late 1990s, AI went into a halt. However, we got the most significant achievement so far; the IBM Computer *Deep Blue* beat the chess grandmaster Garry Kasparov in a highly publicized rematch in 1997. IBM, the American national champion in the production of computers, did not get involved with AI for a long time. Feigenbaum and McCorduck reported that in their book on the *Fifth Generation*, "IBM . . . had a long and—it's irresistible—checkered history of official skepticism about the whole subject of AI. Yorktown Heights, the largest of IBM's research centers, could fairly be described as censorious of, if not downright hostile to, the idea."[1] Part of the Japanese plan was to gain some ground from IBM in an area the company was not strong enough. However, it changed drastically from the late 1990s.

In this period, once again, AI could attract money, people, and everyone's attention because of the revival of neural network-based algorithms that unleashed "machine learning" as the successor to "expert systems."

The story of the factor allowing the machine learning revolution to happen is almost the same everywhere. It is the convergence of massive datasets, mainly because of the internet, increasing computational power, including some innovations such as graphic processing units, also some algorithm innovations.

We may consider Alon Halevy, Peter Norvig, and Fernando Pereira's article *The Unreasonable Effectiveness of Data* as a milestone position on the direction the discipline should go. It is a public position opposed to what expert systems aimed. It is a paradigm shift. The article's title plays with an older article from Eugene Wigner called *The Unreasonable Effectiveness of Mathematics in the Natural Sciences*. Wigner posed the questions of why elegant mathematical formula can explain so much of physics. The conclusion for learning is the opposite, especially when it comes to natural language. Halevy, Norvig, and Pereira say that "we should stop acting as if our goal is to author extremely elegant theories, and instead embrace complexity and make use of the best ally we have: the unreasonable effectiveness of data."[2] It started with an observation of the results regarding a scene completion task in 2008: removing an un-wanted and unsightly automobile from a photograph and filling in the background with pixels taken from a large corpus of other photos. The results were poor with a corpus of thousands of pictures, but the same algorithm performed much better once they had millions of pictures. They argued that would be the case for natural language processing. Their final recommendation was the following,

> Follow the data. Choose a representation that can use unsupervised learning on unlabeled data, which is so much more plentiful than labeled data. Represent all the data with a nonparametric model rather than trying to summarize it with a parametric model, because with very large data sources, the data holds a lot of detail. For natural language applications, trust that human language has already evolved words for the important concepts. See how far you can go by tying together the words that are already there, rather than by inventing new concepts with clusters of words. Now go out and gather some data, and see what it can do.[3]

This method will be widely applied successfully across several problems. It will be the basis for technological *novum*. The applications and consequent scenarios that follow from them will come in the sequence.

In the period, IBM also managed to make another game breakthrough, beat a human champion at the American TV show of questions and answers called *Jeopardy!* The development started in 2005, led by David Ferrucci, the chief scientist of the team developing the *Jeopardy* computer, known as Watson. The victory against Ken Jennings and Brad Rutter occurred in 2011. Ken Jennings, the second highest-earning contestant of all time in the *Jeopardy!* wrote in a final response during the TV show, "I, for one, welcome our new computer overlords." This sentence was said during *The Simpsons'* episode *Deep Space Homer* (1994), affirming that Hal-like computers are the new overlords of the world. As what happened to the chess master Gary Kasparov, Ken Jennings came to the TED Talks stage and told what he had learned from being defeated by a computer.

Regarding AI, he portrayed a side effect of developing this kind of technology, which resembles opinions from people such as the computer scientists Jaron Lanier and writer David Carr[4] who believe people would not learn much more than before. There will be no reasoning memory facts if you access to a perfect search engine all the time. This position resembles one of the comments in the story *The Thought Machine* presented in the Introduction: knowing things by heart given that you have ubiquitous access to all facts on the internet would be as purposeless as trying to dig a deep hole provided you are standing beside a drilling machine. The latest milestone was the Google[5] Deepmind's program AlphaGo's victory over Lee Sedol, a South Korean professional Go player of 9 dan rank, in 2016. Go is a 2,500 years old game, created in Ancient China. It is zero-sum strategy board game. Two players put black and white stones on a grid. The goal is to surround and capture the opponent's stones.

In the cinematographic arena, we have two different groups: before and after 2010s. The first group retells science-fiction ideas from the Asimov period or from classic cyberpunk's period whereas the second group of movies advance the future scenarios of AI and consequently the potential risks of it. The first main public and popular representation of risk of AI in this period is certainly The Wachowskis's *The Matrix* (1999). It creates a dystopian world in which machines dominated the humans after a war. Humans blocked the solar energy as machine's source of power, and machines start to harvest humans as a source of energy. Humans needed to have a mental life, so everyone lives unknowingly inside a simulation of the world in the twenty-first century. We have three kinds of AIs in the movie: the agents, who live in the simulation; the machines and the sentinel, who live in the real world; and the architect, the abstract AI. In the same year, Chris Columbus's *Bicentennial Man* (1999) and Steven Spielberg's *A.I.: Artificial Intelligence* (2001) explore the robots' subjectivity. James Cameron's *Terminator 3: Rise of the Machines* (2003) depicts once again Skynet[6] and the killer robots, and Alex Proyas's *I, Robot* (2004) draws from the Asimov's Robot stories.

Wally Pfister's *Transcendence* (2014) is a film representation of the futurist idea based on Hans Moravec and Ray Kurzweil ideas. From the technological standpoint conceived by them, the movie shows an already existing AI and the following uploading of a human brain to the computer where the superintelligence is. A consequence was the exponential number of innovations. We have already explored both Jonze's *Her* (2013) and *Ex-Machina* (2015). *Ex-Machina* explores the idea already presented in Scott's *Blade Runner* regarding the blurring of the boundaries between intelligent machines and humans to a point that machines question if they are humans. In Hampton Fancher and Michael Green's *Blade Runner 2049* (2017), we will see the same process but reversed for the androids. They started to believe they are humans somehow.

An icon for the film representation is the television series *Black Mirror*, later available on Netflix. The series is a British production created by Charlie Brooker as an anthology of social-science-fiction episodes. The social consequences always depict some side effects. While the series usually depicts some *mindware*, that is, consciousness uploading, consciousness artificially recreated, access to other's memories, and so on; some episodes explore the special application of AI in some products. It is "special" because many AI solutions appear in this series. Still, they are not represented as their main constitutive factor. In the first episode of the second season of Black Mirror in 2013, *Be Right Back*, there is a full-fledge development of an AI chatbot as a commercial service of a synthetic recreation of deceased people. The dead person was recreated using past online communications and social media profiles. Later on the episode, the chatbot could be embedded in a mechanical body. In the episodes of the third season, that appeared in 2016 directly on Netflix, AI is pervasive. The episode *Playtest* explores a horror game driven by a machine learning algorithm incepted in one's brain. So the neural network would progressively learn what would be someone's worse fears. The episode *Hated in the Nation* explores the misuse of autonomous weapons. In the episode, they are a swarm of intelligent drones that can target and terminate someone profiled. In 2017, the autonomous weapons were explored again in the episode *Metalhead*. The representation of this autonomous weapon showed in this episode is inspired by the robot *Spot*, a creation of the company Boston Dynamics. In reality, *Spot* carries objects (up to 30 pounds) in dangerous places. In this series, however, the robot appears to track and kill all the human beings who come across. Again, cultural representations are drawing negative risky scenarios regarding AI that sometimes do not match reality.

Still on autonomous weapons, the short YouTube video *Slaughterbots* related to the Future of Life Institute and Stuart Russell, creates a social-science fiction where swarms of micro-drones are employed to kill any target profiled. The video explores the consequence scenario, especially the drama of a group of students targeted because of political involvement. Stuart Russell narrates at the end of the short video, "This short film is more than just speculation; it shows the results of integrating and miniaturizing technologies that we already have [. . .] AI's potential to benefit humanity is enormous, even in defense, but allowing machines to choose to kill humans will be devastating to our security and freedom."[7] This is an excellent example of how experts act in this new period, using storytelling and fiction to both conceive their scenarios related to AI and to propagate the risk representation of such scenarios for a broad audience by trying to convince it. We will address this phenomenon in the final session of this chapter.

PROPAGATION, POLITICAL, AND INSTITUTIONAL CONSEQUENCES

Drawing from a previous study[8] using data gathered from 3.682.015 public posts containing the term "Artificial Intelligence" on Twitter from January 2007 until January 2018, we could track moments when propagation of risk of perception was accelerated.[9]

At the end of 2014, Stephen Hawking's interview for the BBC about the potential threat of developing human-level AI became viral. His statement was clear, "the development of full AI could spell the end of the human race."[10] Considering the participation of Hawking's interview on the total amount of tweets in the final quarter of 2014 (the period of his interview), it represented 46.5 percent of all risk perception messages and 14.6 percent of all messages related to AI.[11] During this period, James March's *The Theory of Everything* (2014), Hawking's biographic movie, was released. So, the availability bias[12] has probably played its role in the sensibility of public reception regarding Hawking's interview. He drew much credibility from that, since, in the movie, he was portrayed as the highest genius after Einstein, a resilient person and a living example for everyone. After the movie, Hawking went through a broader process of popularization. The availability bias had indeed played a role in the robust amplification process of Hawking's alert because he had expressed his fears about AI in May 2014,[13] when the repercussion was not so high. It represented 29.5 percent of all risk associated messages of AI and 4.9 percent of all messages regarding AI. In this case, the movie industry indirectly affected the prestige related to a specific person who turned out to be a risk communicator, a highly trusted source of information.[14]

These corresponding actions for these kinds of public positions from these experts came from both parts of existing organizations, and they help create new organizations. One of the firsts, the American *Machine Intelligence Research Institute* (2000), pop up to alert the potentially harmful risks of AI. It was followed by the British *Future of Humanity Institute* (2005), based at the University of Oxford, which is concerned with the impact of technology in the reproduction and maintenance of life. Afterward, in the United States, we have Russel's *Future of Life Institute* (2014) supported by Jan Tallinn (Skype's founder). Still, at the University of Cambridge, we find the *Centre for the Studies of Existential Risk* (2012), supported by Elon Musk, and the *Leverhulme Centre for the Future of Intelligence* (2015).

The continuous popularization of AI in the past years was reinforced by economic actions promoted by big corporations, such as incorporations (e.g., Google's incorporation of the AI start-up DeepMind), high investment in in-house products (e.g., IBM's Watson), and the foundation of many start-ups by young people motivated by the euphoria in the area and by billionaires who do not want to be out of this race. That is the example of Elon Musk with his company *Open AI* (2015), which is concerned with AI and its impacts.

The public positioning of these experts, now buttressed by formal organizations, fostered the discussion in other environments, such as politics. In 2016, the Obama's administration conducted a report on AI, developed by a special committee, the *White House's National Science and Technology Council Committee on Artificial Intelligence*.[15] The report brings the first steps of a national plan regarding AI in several areas of society. In 2018, a congresswoman put forward the need of having a national plan regarding AI and the *National Security Commission on Artificial Intelligence*[16] is in charge of it.

In its turn, in UK, *House of Commons* decided to create a special *Science and Technology Committee on Robotics and Artificial Intelligence*[17] (2016) for evaluating the effectiveness of AI's public policies. During the discussion of the committee, experts were called to reading contributions on the topic. Afterward, in 2018, we have another committee in UK, *House of Lords' Committee on Artificial Intelligence*,[18] specially focused on AI. The main aim was to understand to what point UK must invest money in AI in this new twenty-first century AI's race. Another striking example is the creation of a state minister for AI in the UEA.[19]

During this process of politicization of AI, experts have been playing a central role. The *Policy Initiative on AI* launched by the U.S. Obama's government in 2016 was headed by a scholar Edwan Felten, Deputy U.S. Chief Technology Officer, and Professor of Computer Science and Public Affairs at Princeton University. Felten produced the report after a "roadshow" series of workshops promoted by the White House taking place in four prominent universities, during three months, and gathering lawyers, economists, and especially computer scientists. The government made an optimistic approach to AI. It promoted broad and enthusiastic topics of discussion that aimed to foresee the next steps of the technology: "Legal and Governance Implications of AI," "AI for Social Good," "Safety and Control for AI," and "The Social and Economic Implications of AI Technologies in the Near-Term." During Eric Horwitz's talk (a former professor of Computer Sciences at Carnegie Mellon and current director at Microsoft Research Labs) on the context of the safety and control for AI at the third workshop, he revealed the widespread tension from the audience saying that it was as "the elephant in the room." Of course, the "elephant" is the opinion regarding the existential risk AI can bring, and probably the critical motivation for the promotion of the event and the White House report, despite the optimistic and down-to-earth approach to the theme.

Russell's primary intervention was in 2015 in an "open letter" for the *Future of Life Institute* (FLI) written with two other collaborators, Daniel Dewey and Max Tegmark. Signed at present by more than 9,500 people, the open letter embraces many very important intellectuals from different areas of knowledge and great entrepreneurs. The half-life of this particular intervention can be seen below: Tegmark founded the FLI in 2014 to organize initiatives that "mitigate the risk of new technologies can bring." AI is one of these technologies. It is the current main concern and funded area. After the "open letter," the billionaire Elon Musk donated U$10 million to develop global AI research. The goal of the letter, the institute, and Musk's concerns is to develop a "safer" AI.

Since 2014, in addition to the FLI at Harvard, as aforementioned, there are two major centers in Great Britain where the concern for the potential risks of AI are significant, as are their budgets: University of Oxford's *Future of Humanity Institute* (FHI) and the *Centre for the Study of Existential Risk* (CSER) at the University of Cambridge. Finally, the center was founded by Russell at the University of Berkeley in 2016, the *Centre for Human-Compatible Artificial Intelligence* (CHCAI). All the important names responsible for forming this broad and recent insight into the potential negative impacts that AI can bring to society gravitate around these centers, which are not by any chance closely related to some of the world's top universities. For instance, the catastrophes' philosopher Nick Bostrom is the head of *FHI* and the author of the most influential book on the risks of AI: S*uperintelligence: Paths, Dangers, Strategies* (2014). At the *CSER*, Stephen Hawking was a vital member of the council, responsible for the public commentaries spread through media that shaped the actual risk perception of the technology. Especially regarding the irreversible risk that the development of a much more powerful intelligence than the human could bring.

The influence of these intellectuals and their centers on academia, media, and, consequently, the general public cannot be ignored, especially during the current economic time when large technology companies are in an "arms race" for the development of AI.

Many current topics of innovation involve AI projects. The opinion of computer scientists, such as Russell, has become highly valued, to the extent that he has been one of the speakers at the World Economic Forum since 2015. It is a time when computer scientists came out of their labs to the forefront of TED Talks. They definitely have become rock stars. Finally, the competition has extended to countries whose governments devise their national plans, such as the United States, the United Kingdom, and China— in particular, the Chinese *Next-Generation Artificial Intelligence*

Development Plan explicitly states the nation's goal of world domination in the technology by 2030.

EXPERTS GETTING ON THE STAGE: THE PROLIFERATION OF TED TALKS

TED Talks started in the 1980s, but it skyrocketed in the 2000s when the British entrepreneur Chris Anderson bought it and became the curator of the event. Then, the event became a global phenomenon. TED stands for *Technology, Entertainment, and Design*, a name that would illustrate the idea of convergence of these three activities. In a TED talk, the speaker aims to spread knowledge using storytelling. The talks are populated by personal anecdotes to create an empathetic link, thus a credibility link with the audience. Credibility is the driving force for knowledge dissemination. As we have been suggesting, these kinds of online shared contents have allowed experts to achieve direct contact with laypeople and communicating to them their images of the consequences of technological innovation directly. In this regard, TED, as the most successful venue for sharing knowledge content, will be taken as legit by laypeople. The old and typical triangularization "scientist-mediator-audience" is finally over with TED's direct format in which a scientist is in a stage talking directly with people.[20] Besides that, TED expresses the trend we saw with Moravec and Kurzweil in the past chapter, namely, that an expert could conceive scenarios outside a fictional narrative. This kind of talk converges personal stories with factual scientific knowledge, but it is often caught in the job of thinking about the future (possible) consequences of the scientific invention. We can also suggest that experts' main concerns are future social implications whenever they address the future impacts of a scientific discovery or a technological innovation. It is counterfactual thinking, or a social-science-fiction style of thinking, but freed from the explicit fictional work. As a consequence, one can conceive TED Talks and similar talks as a hybrid and emerging genre.[21]

The primary purpose of these talks is to get the most views, likes, shares, and comments as possible. So, it is a contemporary entertainment business. It differs significantly from, for example, Norbert Wiener's call in *The Human Uses of Human Beings* in which selling the book as much as possible was not the primary goal of his undertaking. This kind of effort belongs to the traditional university press business model, aiming a niche audience for selling books. From our contemporary standpoint, we could easily convert Wiener's and Dominique Dubarle's critique (see chapter 2) in TED Talks. Contents for this kind of venue have always existed; the difference lies in how the messages are conveyed. A succeeded talk depends on the speakers' performance, that is, how they are listened, remembered, and perceived as "credible promoters of knowledge dissemination."[22] Conversely, it also shapes the way experts promote research and scholarship.[23] Practically, any expert could convey their research into talk of this kind. So it would affect how scholarship has been produced—at least how the results are presented.

This emergent genre is not free from criticism despite its response to the audience's expectative of the rhetoric effect in the digital media. Kedrowicz and Taylor devised an experiment where students responded to two questionnaires in two different classes: a basic public speaking course and a technical communication course. They concluded from the experiment results that students' perception of public speaking is now related to "a coproduced, networked, and engaging narrative."[24] Back to 2013, Benjamin Bratton, an associate professor of visual arts and director of the Center for Design and Geopolitics at the University of California, wrote a harsh public criticism about the TED-genre for the British newspaper *The Guardian*. There he claims that "Science, philosophy and technology run on the model of American Idol—as embodied by TED Talks—is a recipe for civilizational disaster."[25] The oversimplification is at the heart of this issue.

Regardless of the criticism on the TED-genre, the need for engagement and the oversimplification of ideas allow the speaker to talk about the chain of future events in which some links are just too unlikely to occur. The audience's perception of these narratives fills them with affective and availability biases. In this sense, because speakers engage the audience with personal anecdotes and untamed conjectures, they may conceive their chain of scenarios without analyzing their biases.

Futurists have their guaranteed place in the conference. For instance, Ray Kurzweil gave three talks (posted on the official website in November 2006, June 2009, and June 2014) on his ideas of singularity. Another philosopher-futurist, Nick Bostrom, appeared in three opportunities to talk about *existential risks* (July 2005, March 2015, April 2019), including AI. In Bostrom's 2015 talk,[26] he mentioned a survey he conducted with AI experts and their guesses about when we will have a superintelligent machine. The result indicated optimism from the AI community in achieving its goal by the mid of this century (twenty-first). In this talk, he reproduces some of the arguments of his best-seller book,[27] especially that superintelligence may be the last invention of humankind. It would happen chiefly because intelligence would be equivalent to optimization, and that would lead us to a King Midas problem—broadly speaking, for this case, a story for the tragic problem of strict rationality. This position converges Stuart Russell's argument about the need for value alignment between AI and humans in his book *Human Compatible*. In 2017, Russell also gave a TED talk[28] in which he pointed out the King

Midas problem and the necessity of goal and values alignment between humans and computers. He also conceives imagined situations to illustrate the hypothetical risk of human-machine values' misalignment.

> So imagine your domestic robot's at home. You're late from work again and the robot has to feed the kids, and the kids are hungry and there's nothing in the fridge. And the robot sees the cat. (Laughter) And the robot hasn't quite learned the human value function properly, so it doesn't understand the sentimental value of the cat outweighs the nutritional value of the cat. (Laughter) So then what happens? Well, it happens like this: "Deranged robot cooks kitty for family dinner." That one incident would be the end of the domestic robot industry. So there's a huge incentive to get this right long before we reach superintelligent machines.[29]

We can see from Russell's example of how the scenarios are conceived and communicated. We start with the technological *novum*, the invention of a robot that can not only cook but also hunt (a domestic cat). This general robot hunts the cat and prepares it for dinner. It knows not only how to evaluate the nutritional value of the cat but also how to clean and prepare dishes made from cats. Finally, this imagined damage has its secondary social and economic effects. Here, it would end the domestic robot industry, which already existed in his example. It is unclear how this chaining of events could happen. This sequence has the faintest possibility to happen, especially if we unpack every step to consider. So, what is this kind of scenario for? Indeed, it serves a communicative, rhetoric effect. It has its *true* in the imaginary world—as we saw with Lewis.[30] A scenario like this one takes place into an entirely imaginary world, but it is not announced as fictional as a science-fiction world, so the storyteller never asked the suspension of disbelief for the audience.

The physicist and co-founder of the *FLI* Max Tegmark also argues that the real worries we must have regarding AI are due to the human-machine goals misalignment. He expresses that in his book *Life 3.0*[31] as well as in this TED talk.[32] Like other authors we have seen, Tegmark disregard scenarios where AI became conscious or became evil, dystopian, as mythical. However, this more practical representation of the risk of AI is not free from imagined scenarios.

Back to the catastrophic scenarios, we can see that future existential risks related to AI were still portrayed. Existential risks are, in short, any risk that may bring global catastrophe or human extinction, for instance, asteroid collision, nuclear warfare, a pandemic, and so on.[33] The astrophysicist Martin Rees also talked about existential risk on two occasions (January 2007 and August 2014). On the second occasion, he mentions future risk related to AI: dumb robots going rogue (a classic risk as we have seen over the past century) or network developing a mind (a risk imagined with the cyberpunk science fiction). Bill Joy contributed in February 2006, drawing from his article published on Wired, which we have already analyzed in the previous chapter. The philosopher Sam Harris gave another talk about the general risk of AI, but his talk went viral with 5 million views on the ted.com.[34] He talked about I.J. Good's idea of an intelligence explosion and also rose the risk of annihilation if there were goals and values' misalignment between machines and humans. He ties to the representation of the risk of AI an emblematic name from the nuclear weapons, the Manhattan Project, "I think we need something like a Manhattan Project on the topic of artificial intelligence. Not to build it, because I think we'll inevitably do that, but to understand how to avoid an arms race and to build it in a way that is aligned with our interests."[35] Before this talk, Sam Harris gave an even more popular talk[36] with more than six million views on ted.com about how science could answer moral questions. He claimed that science should be an authority on moral issues; consequently, it should shape human values. It is also interesting to see how futurists became an official label to designate some of the speakers.

Besides that, optimists futurists abound. Experts that follow futurists' ideas of mind uploading or turning into cyborgs such as Kurzweil's. For instance, the technologist Arnav Kapur[37] argued for AI as a mind extension in 2019. Two years earlier, the sociologist and futurist Robin Hanson, also talked about mind uploading and crafted a "*the thought machine*" like narrative. Hanson creates a (non) social-science fiction in his book *The Age of Em*[38] in which he explores the consequences of the new technology, large-scale production of smart robots with brain emulations or ems, as he calls it. In April 2017, the co-creator of Siri, Tom Gruber, talks about memory enhancement and the collaboration between humans and AI.[39] Those who stand for coevolution between human beings and machines are—almost by definition—optimists. The best examples we addressed in the previous chapter were Hans Moravec and Ray Kurzweil's positions. In 2012, there was also a talk regarding Shyam Sankar[40] argued for the human-computer close cooperation and Licklider's argument (see chapter 4) for human-computer symbiosis.

We still have future risks conceived in dystopian scenarios. For example, a Legal scholar, Nita Farahany,[41] claimed in a TED Salon (a local event) that we will have brain decoding technology soon based on the combinations of electroencephalogram (EEG) and machine learning. For these future scenarios, she pointed out two movies Jane Weinstock's *The Moment* (2013) on false memories and Steven Spilger's *Minory Report* (2002), which is based on Phillip K. Dick's short story *The Minority Report* (1956). Up to the end of 2019, there were twenty-two futurists.

We use the term *futurism* in association with the twentieth-century artistic movement that rejected traditional forms and welcomed the dynamism of modern technology. A futurist may be an adherent of this movement, which is the Oxford dictionary's first definition. In a more abstract sense, according to the Oxford dictionary, a futurist is "a person who studies the future and makes predictions about it based on current trends,"[42] which is closely associated with *futurology* as "systematic forecasting of the future, especially from present trends in society."[43] Futurology is a term also related to the job of fortune-tellers, seers, soothsayers, diviners, and prophets as well. From a Weberian classic sociological approach, these classes of people are considered to bear charisma. Charisma is "the extraordinary," in Weber's words, "it can be a gift bound to the object or person by nature, which cannot be acquired. Either, it may and must be provided to the object or person artificially by certain extraordinary means."[44] The association between telling something about the future with the role of religion is hard to avoid. Even the third definition of a futurist in the Oxford dictionary brings this association, "a person who believes that eschatological prophecies are still to be fulfilled."[45]

The TED Talks may also contribute to increasing the availability bias on the public perception of any topic. In our case, it increased the public perception that AI (the full-fledged technology) is imminently tied to the future risks it may bring.

As we saw with the Twitter history of posts, the TED Talks about AI had its momentum between 2016 and 2018, sixty talks out of ninety-five. Before 2013, the term was not widely used. Even at the beginning of 2013, Andre McAfee presented a talk on the idea of robots taking our jobs and societal impacts of that.[46] His talk derived from a previous TED talk in 2012 whose title was *Are Droids Taking Our Jobs?*[47] He employed the term droid instead of robots or AI. Across the new peak, it will change. The term AI will work as an even stronger magnet for all dispersed terms. The raised flag of the threats of unemployment driven by AI is expected. He has been a scenario conceived since *Ammianus' The Thought Machine* presented in the introduction. The risk of human work being superseded by AI is always around the corner. Martin Ford is the other futurist who has been working with the impacts of AI in work. In his TED talk, we mention the false alarms in the past, but after the example of AlphaGO, we may wonder if it is impossible to have machines doing jobs we previously thought were safeguarded. He conceives the future scenarios in which we will have an economic problem from income distribution and ways to overcome them, such as the basic income policy.

We can say that the risk from misalignment of values belongs to the category of future AI risks. In the last years, present risks are mainly related to privacy risks, including corporations and governmental exploitation and biased algorithms. The TED conferences became a platform for communicating the message of this kind of risk associated with real situations. One of the most influential critics of the misuse of AI and other modern services, the Virtual Reality, Jaron Lanier,[48] calls attention alienation. His criticism is aligned with some of previous cyberpunk criticism, and alienation is the most important. In the late 2000s, in an article for the edge.org., he criticized how we start to believe in the internet as an entity and how it would have agency. These sorts of beliefs devalue "those people [creating the content] and making ourselves into idiots,"[49] an argument further developed for years later in his famous book *You Are Not a Gadget.*[50] Other talks about the economic exploitation of data for training machine learning models to targeting and profiling, for example, the technology and investigative journalists Kashmir Hill & Surya Mattu in 2018, the sociologist, Zeynep Tufekci, in one of her talks about the techno-sociologist Dystopia from AI, that is, companies use AI for controlling us entailing in potential, significant risk. In 2018, data protection and anti-spying devices were also addressed by the Information Security Researcher, Finn Lutzow-Holm Myrstad, and the data privacy expert, Derek Banta.

Biased algorithms due to design or databases rose as a real present concern. As we mentioned, it became part of the representations of risks of AI in machine learning, especially with the neural network approach. It was never portrayed as a future risk before. In fact, the perception of this kind of risk is even more recent. Taking the TED as a thermometer, in 2013, we can see criminal justice reformer, Anne Milgram,[51] praising the application of statistics to crime without thinking about the side effects. In the early 2010s, the idea of data-drive or solely big data was not easily combined with the concept of machine learning and later AI. She makes this public defense of data-driven police and a universal risk assessment tool, provided that the closest representation in entertainment so far as *Robocop* and Minority Reports, that is, alternative worlds separated by a vast gap from our world.

The governmental usage of AI is now at the core of many representations of AI's future risks. In the Weiner-Dubable debate, we saw how Dubable addressed this point. However, it was entirely hypothetical in that case. During the 2010s, we got pieces of evidence from how governments are doing that. The first and most emblematic case was the NSA's case denounced by Snowden. However, it was not associated with AI and automated decisions. So, the core example in this decade became China's government. This point was addressed in a TED Talk by Gary Liu, a media executive in 2018.

Regarding the biases, besides the bias on incarceration's judgments, big tech company's applications caused moral harm. This is conveyed by the documentarist Robin Hauser regarding the implicit Bias in AI and the cases of the Microsoft Twitter Bot and Google's classification of gorillas. In her first two talks, the sociologists, Zeynep Tufekci, came for the idea that machine learning requires moral behavior. The mathematician also spoke

Cathy O'Neil about this problem, drawing from her book "weapons of math destruction."

A new imminent threat is under consideration, the *deep fakes*. Up to this point, there was no popular scenario where AI could be harmful that included that. It became something possible in the proliferation of applications of deep learning techniques in the late 2010s. The computer scientist Supasorn Suwajanakorn created fake videos and considered at the same time their potential misuse. As a societal impact from the future risk, he created a foundation called reality defender. Doug Roble, an expert in Computer Graphics, presented in a TED talk about how deep learning can lead to deep fakes and their consequences. Some experts have thought about the following conceived scenarios. For instance, law professor Danielle Citron talked about how deep fakes could undermine democracy, and cause social hazard via pornography and even financial markets in July 2019. Deep fakes have the power to blur the real and artificial facts.

Besides the moral and societal hazard from AI, a representation of the risk of AI that could bring physical harm appeared with the possibility of smart weapons. These representations appeared before. We also addressed the recent secondary social impacts from the representation of the risk of AI weapons with the initiative, such as the Ban on smart weapons.

It has been portrayed in the image of the army of robots since the beginning of science fiction as we saw, and it also appeared regularly in games.[52]

The risks of using AI in warfare appeared in TED already in 2009. The former mariner, PW Singer, talks about the consequences of using robots in war because it changes the responsibility of decisions.

The contemporary science-fiction writer Daniel Suarez further explores the fiction to the real possibility in the case of employing autonomous weapons in warfare and societal control. He claimed that the kill decision should not belong to a robot. Among the many experts and speakers, we presented so far in this session; Daniel Suarez is the specialist in creating future scenarios. So, he gives in his talk part of the consequences from the introduction of this technological *novum.*

> You might be thinking that citizens of high-tech nations would have the advantage in any robotic war, . . . but I think the truth is the exact opposite. I think citizens of high-tech societies are more vulnerable to robotic weapons, and the reason can be summed up in one word: data. Data powers high-tech societies . . . we are more visible to machines than any people in history, and this perfectly suits the targeting needs of autonomous weapons. What you're looking at here is a link analysis map of a social group. . . . Now, if you're a marketer, you might target them with product samples. . . . But if you're a repressive government searching for political enemies, you might instead remove them . . . disrupt their

> social group. . . . Now in a world of cheap, proliferating robotic weapons . . . popular movements agitating for change could be detected early, and their leaders eliminated before their ideas achieve critical mass. . . . Anonymous lethal weapons could make lethal action an easy choice for all sorts of competing interests. And this would put a chill on free speech and popular political action, the very heart of democracy . . . we need an international treaty on robotic weapons and, in particular, a *global ban* on the development and deployment of killer robots . . . robotic weapons might be every bit as dangerous, because they will almost certainly be used, and they would also be corrosive to our democratic institutions.[53]

Besides all the representations of risks of AI, the reference to artificial is usually a tool that can enhance some ongoing process or help in a specific task. For example, Jack Dorsey, Twitter's CEO, told a TED interview (2019) how they applied machine learning and deep learning to detect tweets of harassment proactively. AI devices for health are a frequent topic—for example, Rosalind Picard (2018), Poppy Crum (2018). The expert Fei-Fei Li (2015), a leading researcher in the ImageNet project, also mentioned the health narrative for helping the future. Joel Selanikio (2013), a Heath and tech activist, talked about how Big Data in health as an optimized way of collecting and processing public health data. The famous roboticist Rodney Brooks (2013), talked about the need for developing AI robots chiefly because of aging.

In response to the threat of superseding, we start to observe attempts to shift the representation of AI as a collaborator, a human productivity enhancer. The music entrepreneur, Pierre Barreau (2018), also see the beneficial use of AI in personalized composing music. He takes Spike Jonze's *Her* as a key example. Kasparov (2017) talked about human-machine cooperation instead of replacement—the idea that the triumph of machines is a triumph of humans. In the world of investments, the manager of the largest hedge fund in the world, Ray Dalio (2017), argued for working with AI. Collective decision-making (humans and algorithms) is better than individual decision-making.

Still, in the arguments for the cooperation between humans and AI, Erik Brynjolfsson (2013) came with the idea of the economy of abundance. Kevin Kelly (2016), the founder of Wired, talked about how AI can produce a second industrial revolution since it can be applied in the background of many things. (Mechanical Engineer BCG's) Markus Lorenz (2015) Intelligence machines will bring Industry 4.0; optimist regarding jobs.

As we said before, this is TED-like genre in the way we create scenarios are not confined to the TED Talks. We can see this argument in books as well. For example, the former expert and executive Kai-Fu Lee depicted a perfect example of what we have been arguing in his latest book, *AI Superpowers.*[54] At some point, he came with the example of the "Yonghui Superstore."

> It's always a nice feeling when your shopping cart greets you like an old friend. As I pull the cart back from the rack, visual sensors . . . completed a scan of my face and matched it to a rich, AI-driven profile of my habits, as a foodie, a shopper, and a husband to a fantastic cook of Chinese food. . . . "Let me know if there's anything you'd like to add or subtract from the list," the cart chimes in. "Based on what's in your cart and your fridge at home, it looks like your diet will be short on fiber this week. Shall I add a bag of almonds or ingredients for a split-pea soup to correct that? . . . The cart is speaking in Mandarin, but in the synthesized voice of my favorite actress, Jennifer Lawrence. . . . The cart moves autonomously through the store, staying a few steps ahead of me. . . . The precise location and presentation of every item has been optimized based on perception and purchase data gathered at the store. . . . All the concierges are knowledgeable, friendly, and trained in the art of the upsell. It's far more socially engaged work than traditional supermarket jobs, with all employees ready to discuss recipes, farm-to-table sourcing, and how each product compares with what I've tried in the past.[55]

He concludes from this imagined AI-driven supermarket that,

> Perception AI–powered shopping trips like this will capture one of the fundamental contradictions of the AI age before us: it will feel both completely ordinary and totally revolutionary. Much of our daily activity will still follow our everyday established patterns, but the digitization of the world will eliminate common points of friction and tailor services to each individual. . . . A supermarket like the one I've described isn't far off. The core technologies already exist, and it's largely a matter now of working out minor kinks in the software, integrating the back end of the supply chain, and building out the stores themselves.[56]

Kai-fu Lee's idyllic consumer experience reveals how AI can enable and enhance any technological sophistication, provided we ignore technical, economic, and conceptual difficulties in the development of technology. It has been progressively becoming more amorphous, more abstract, hence ubiquitous. The technology was intellectualized in the same sense that occurred in the biblical god. It began its career historically probably as a sort Baal and progressively it was conceptualized in the most abstract form, a God that is portrayed as a general intelligence of the universe even for non-Christians. Regarding AI, it started as its embodied version, as a concrete machine intelligence in robots as units. It has been progressively abstract in a system-like intelligence that can also contain both particular machine intelligence, such as specialized robots, as well as human intelligence. All of these pieces would be part of an intricate system.

As a final remark, an interview with Daniel Hillis, an expert from the second generation of AI researchers, and student and friend of Marvin Minsky, illustrates perfectly this process with his personal changing of belief.

> I've changed my idea of what I think artificial intelligence is going to be like. I think when I first started thinking about it, I kind of imagined the science fiction I Robot of an embodied artificial intelligence of being very much like a human, but made of electronic parts or something like that. And it's not that I think that's impossible to build, but I don't think that's the interesting thing. . . . And so I think we fall into the trap of imagining an artificial intelligence as being too analogous to human intelligence. So people ask what seem to me silly questions sometimes, like they say, "Well, there's a self-driving vehicle. Would it be willing to sacrifice the driver to prevent running over a pedestrian?" As if it's going to have a little brain in it like the driver does. But actually, that decision is going to be made the system, by society, by the laws we set, by all the experience of all the self-driving vehicles. And so in some sense, that's a much better way of making a decision than having the driver sitting in the driver's seat make that decision. It's going to be a much more informed decision that a lot more participate in and is done for much less selfish reasons. So because that decision is going to be made by the system of all the artificially intelligent vehicles, not by a single individual self-preserving AI brain. So I now think of intelligence as kind of extended intelligence. It's the intelligence of all the machines working together and all the people working with them. So it's the kind of intelligence, in some sense, like is in Google right now. Google isn't really machine intelligence, it's not really human intelligence, it's some combination of humans and machines that are able to find things that you're searching for. And I think AI is going to be like that. It's going to be a combination of machines and people and so it's that extended group of intelligences that's going to be the thing that does the thinking.[57]

NOTES

1. Feigenbaum and McCorduck, *The Fifth Generation*, 178.
2. Alon Halevyet al., "The Unreasonable Effectiveness of Data," *IEEE Intelligent Systems*, (April 2009), 8.
3. Halevy et al., "Unreasonable Effectiveness of Data," 12.

4. See Nicholas Carr, *The Shallows: What the Internet is Doing to our Brains* (New York: W.W. Norton & Company, 2010).

5. See also the Google Brain Project and the development of the AutoML. Judith Zamith-Cruz and Paulo Vieira, "Can Machine Intelligence Overcome the Human One?" in *Automata's Inner Movie: Science and Philosophy of Mind* ed. Manuel Curado and Steven Gouveia (Wilminton: Vernon Press, 2019).

6. Skynet is a fictional artificial neural network built by Cyberdyne Systems that serves as the main antagonist of the Terminator franchise. It was based on group mind and artificial general intelligence system.

7. Stewart Sugg, *Slaughterbots*, Space Digital, 2017, https://www.youtube.com/watch?v=9CO6M2HsoIA.

8. This study is part of a postdoc researcher undertook at the University of Cambridge under the supervision of Patrick Baert.

9. For further analysis of this dataset, see Neri and Cozman, "The Role of Experts."

10. Hawking, "Stephen Hawking Warns Artificial."

11. Neri and Cozman, "The Role of Experts."

12. See chapter 1 of this book, Tversky and Kahneman, "Judgment under Uncertainty: Heuristics and Biases," and Kahneman, Slovic, and Tversky, Judgment under Uncertainty: Heuristics and Biases.

13. Stephen Hawking, Stephen Hawking is Terrified of Artificial Intelligence, interview by C. Kolodny, *Huffington Post*, May 5, 2014, https://www.huffingtonpost.co.uk/entry /stephen-hawking-artificial-intelligence_n_5267481.

14. Neri and Cozman, "The Role of Experts."

15. NSTC, "National Science and Technology Council Committee on Technology. Preparing for the Future of Artificial Intelligence" (NSTC—Executive Office of the President National Science and Technology Council, 2016), https://obamawhitehouse.archives.gov/sites/default/files/whitehouse_files/microsites/ostp/NSTC/preparing_for_the_future_of_ai.pdf.

16. Elise Stefanik, "H.R. 5356 to Establish the National Security Commission on Artificial Intelligence," 2018, https://www.congress.gov/bill/115th-congress/house-bill/5356.

17. House of Commons, "Robotics and Artificial Intelligence," House of Commons, 2016, https://publications.parliament.uk/pa/cm201617/cmselect/cmsctech/145/14502.htm.

18. House of Lords, "AI in the UK: Ready, Willing and Able?" House of Lords, 2018, https://www.parliament.uk/business/committees/committees-a-z/ lords-select/ai-committee/news-parliament-2017/ai-report-published/.

19. D. Galeon, "Dubai Just Appointed a "State Minister for Artificial Intelligence," *Futurism*, October 20, 2017, https://futurism.com/dubai-just-appointed-a-state-minister-for-artificial-intelligence.

20. Giuseppina Scotto di Carlo, "New Trends in Knowledge Dissemination: TED Talks," *Acta Scientiarum. Language and Culture* 36, no. 2 (May 9, 2014), 121, https://doi.org/10.4025/actascilangcult.v36i2.22619.

21. Julia Ludewig, "TED Talks as an Emergent Genre," *CLCWeb: Comparative Literature and Culture* 19, no. 1 (March 1, 2017), https://doi.org/10.7771/1481-4374.2946.

22. Giuseppina Scotto di Carlo, "Ethos in TED Talks: The role of Credibility in Popularised Texts," *Facta Universitatis*, Linguistics and Literature, 12, no. 2 (2014), 81–91.

23. Antonio Compagnone, "The Reconceptualization of Academic Discourse as a Professional Practice in the Digital Age: A Critical Genre Analysis of TED Talks," *HERMES - Journal of Language and Communication in Business* 27, no. 54 (December 22, 2015), 49, https://doi.org/10.7146/hjlcb.v27i54.22947.

24. April A. Kedrowicz and Julie L. Taylor, "Shifting Rhetorical Norms and Electronic Eloquence: TED Talks as Formal Presentations," *Journal of Business and Technical Communication* 30, no. 3 (July 2016), 352–377, https://doi.org/10.1177/1050651916636373.

25. Benjamin Bretton, "We Need to Talk About," *The Guardian*, December 30, 2013, http://www.theguardian.com/commentisfree/2013/dec/30/we-need-to-talk-about-ted.

26. Nick Bostrom, "What Happens When Our Computers Get Smarter than We Are?" March 2015, https://www.ted.com/talks/nick_bostrom_what_happens_when_our_computers_get_smarter_than_we_are.

27. Nick Bostrom, *Superintelligence: Paths, Dangers, Strategies* (Oxford: Oxford University Press, 2014).

28. Stuart Russell, "3 Principles for Creating Safer AI" April 2017, https://www.ted.com/talks/stuart_russell_3_principles_for_creating_safer_ai.

29. Russell, "Principles for Creating."

30. Lewis, "Truth in Fiction."

31. Max Tegmark, *Life 3.0: Being Human in the Age of Artificial Intelligence* (New York: Knopf, 2017)

32. Max Tegmark, "Max Tegmark: How to Get Empowered, Not Overpowered, by AI," April 2018, https://www.ted.com/talks/max_tegmark_how_to_get_empowered_not_overpowered_by_ai.

33. Martin Rees, *Our Final Hour: A Scientists Warning; How Terror, Error, and Environmental Disaster Threaten Humankind's Future in This Century—On Earth and Beyond* (New York: Basic Books, 2003).

34. Sam Harris, "Can We Build Ai without Losing Control over It?" June 2016, https://www.ted.com/talks/sam_harris_can_we_build_ai_without_losing_control_over_it.

35. Harris, "Can We Build Ai"

36. Sam Harris, "Science Can Answer Moral Questions," February 2010, https://www.ted.com/talks/sam_harris_science_can_answer_moral_questions.

37. Arnav Kapur, "How AI Could Become an Extension of Your Mind," April 2019, https://www.ted.com/talks/arnav_kapur_how_ai_could_become_an_extension_of_your_mind.

38. Robin Hanson, *The Age of Em: Work, Love and Life When Robots Rule the Earth* (Oxford: Oxford University Press, 2016).

39. Tom Gruber, "Https://Www.Ted.Com/Talks/Tom_gruber_how_ai_can_enhance_our_memory_work_and_social_lives," April 2017.

40. Shyam Sankar, "The Rise of Human-Computer Cooperation," June 2012, https://www.ted.com/talks/shyam_sankar_the_rise_of_human_computer_cooperation.

41. Nita Farahany, "When Technology Can Read Minds, How We Will Protect Our Privacy?" November 2018, https://www.ted.com/talks/nita_farahany_when_technology_can_read_minds_how_will_we_protect_our_privacy.

42. Oxford Dictionary, "Futurist," in *Oxford Dictionary of English* (Oxford: Oxford University Press, 2019).

43. Oxford Dictionary, "Futurology," in *Oxford Dictionary of English* (Oxford: Oxford University Press, 2019).

44. Max Weber, *Max Weber-Gesamtausgabe (MWG): Wirtschaft Und Gesellschaft. Die Wirtschaft Und Die Gesellschaftlichen Ordnung Und Mächte. Nachlaß. Religiöse Gemeinschaften.*, ed. Hans Kippenberg and Petra Schilm, vol. 22–2, Max Weber-Gesamt Ausgabe, I (Tübingen: Mohr Siebeck, 2001), 122. For a detailed account of the role of experts in the socio-historical process of rationalization and intellectualization, see Hugo Neri, "The Process of Intellectualization: foundations for a sociological explanation of knowledge" (University of São Paulo, 2014), https://teses.usp.br/teses/disponiveis/8/8132/tde-13082014-210252/publico/2014_HugoNeriMunhoz_VOrig.pdf.

45. Oxford Dictionary, "Futurist."

46. Andrew McAfee, "What Will Future Jobs Look Like?" February 2013, https://www.ted.com/talks/andrew_mcafee_what_will_future_jobs_look_like/transcript.

47. Andrew McAfee, "Are Droids Taking Our Jobs?" June 2012, https://www.ted.com/talks/andrew_mcafee_are_droids_taking_our_jobs.

48. Jaron Lanier, "How We Need to Remake the Internet," April 2018, https://www.ted.com/talks/jaron_lanier_how_we_need_to_remake_the_internet.

49. Jaron Lanier, "DIGITAL MAOISM: The Hazards of the New Online Collectivism," *Edge.Org*, May 29, 2006, https://www.edge.org/conversation/digital-maoism-the-hazards-of-the-new-online-collectivism.

50. Jaron Lanier, *You Are Not a Gadget: A Manifesto* (New York: Vintage Books, 2010)

51. Anne Milgram, "Why Smart Statistics Are the Key to Fighting Crime," October 2013, https://www.ted.com/talks/anne_milgram_why_smart_statistics_are_the_key_to_fighting_crime.

52. Even though we did explore that influence of games on the representation of artificial intelligence, which is probably relevant. As a personal anecdote, the familiarity with the term artificial intelligence came from games for me, primarily related to the level the computer game playing.

53. Daniel Suarez, "The Kill Decision Shouldn't Belong to a Robot," June 2013, https://www.ted.com/talks/daniel_suarez_the_kill_decision_shouldn_t_belong_to_a_robot.

54. Kai-Fu Lee, *AI Superpowers: China, Silicon Valley, and the New World Order* (Boston: Houghton Mifflin Harcourt, 2018).

55. Lee, *AI Superpowers: China.*

56. Lee, *AI Superpowers: China*, epub.

57. Daniel Hillis, interview by Christopher Sykes and George Dyson, October 2016, https://www.webofstories.com/play/danny.hillis/1.

Chapter 6

Let Us Understand Our Relationship with Artificial Intelligence (in Collaboration with Veridiana Domingos Cordeiro)

Heretofore, we have presented throughout the chapters the role of experts in conceiving scenarios in which AI could harm humans. We exposed how experts and fiction writers operate similarly whenever it comes to the counterfactual building of the further consequences of any technology. Unsurprisingly, many scenarios containing future risks of AI had been thought before the expert's depiction of them. It is also true that the boundaries of fictionality of these scenarios became blurred. Regardless of the plausibility of the scenarios, we restricted ourselves in this book to present the progressive blurring of the roles (both the futurist and the expert) as soon as the experts became more autonomous. Furthermore, we tried to present how future risk scenarios came to be as well as their secondary impacts. We also restricted ourselves to a higher level when it comes to the details of the historical events; otherwise, every single historic event would be a book in its own merit. Besides that, ee focused only on a Western "history" for AI. An Eastern history for AI also must be conceived. To ignite this spark, in this chapter, we will draw some crosswise reflections on how eastern culture (we will talk especially about Japan) understand and relate with AI.

While the relation between Western societies and AI pointed to a negative direction, Japan's culture did otherwise. The tripod "cultural representation, media, and experts" that affects the social amplification of risk varies socially and culturally. This trend seems to be historically sedimented. In Japan, machine intelligence is not only positively depicted in movies, mangas, and *animés*, but also Japanese experts do the same, thus enhancing a positive bias with images of familiarity regarding the artificial entities.

We also must add here that the positive influences are not only made of "symbolism," but also the socioeconomic situation propels the positive or the negative direction of the influence. The Luddite feelings in Western countries due to the industrialization may persist, especially in economies with imminent unemployment because of the machines. As we have seen, applied AI means human substitutions in some tasks.

THE AI CIRCLE: EMOTIONAL SIGNS

As an exercise to understand how this cycle is a generator matrix for the AI representations within society in general, we have proposed to people to draw

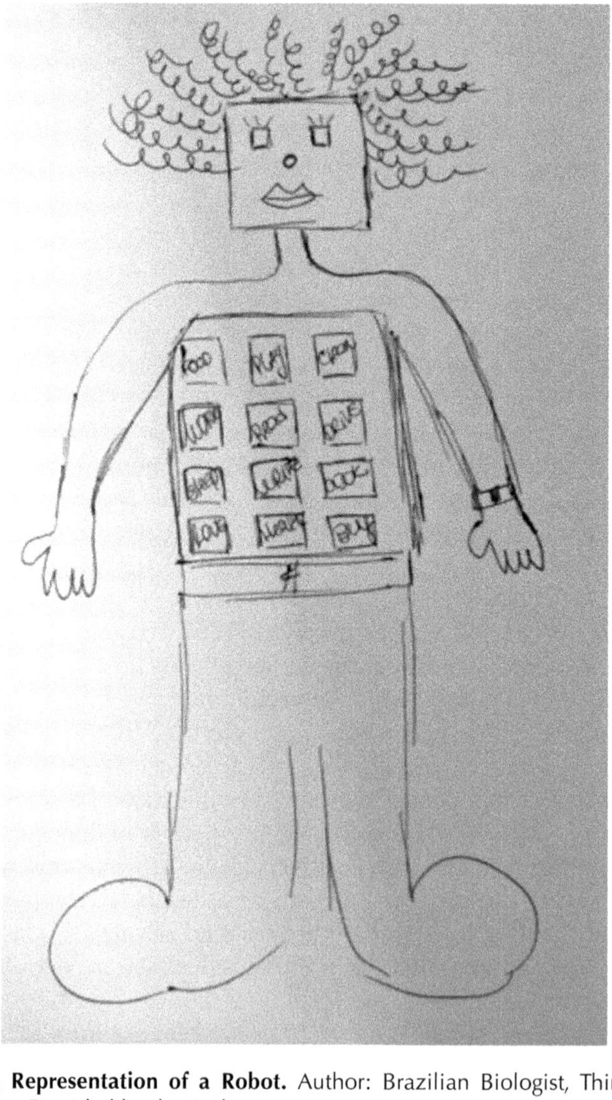

Figure 6.1 A Representation of a Robot. Author: Brazilian Biologist, Thirty-Siz Years, Woman. Image Provided by the Author.

what comes to their minds when they think of AI. Some questioned if I was thinking about robots, and some only freely draw it. By observing the images, we make sense of how people represented sparse ideas related to AI. Of course, we know that this exercise does not match an experimental criteria, so we using it just to raise some insights.

The robot (Figure 6.1) is a female representation drawn by a woman. She has fuzzy hair and an outer lip. According to the drawer, the buttons on her belly represent full command skills that the robot can perform, such as buying, driving, reading, playing, working, and writing. However, she seems to also have emotional and physiological skills, which are commonly inherent to humans, such as loving.

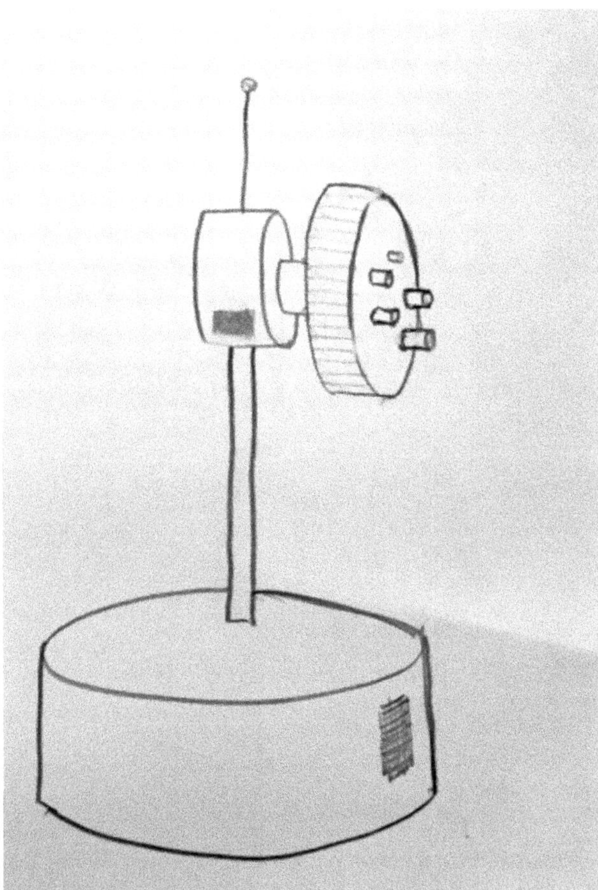

Figure 6.2 A Representation of a Robot. Author: Brazilian Artist, ThirtyYears, Man. Image Provided by the Author.

The two robots (Figure 6.2 and Figure 6.3) look like old fashioned machines. Both are static because they do not have wheels or any other possibility to move around. These two representations diverge from the others that have anthropomorphic features.

The robot (Figure 6.4) does not have wheels; however, it has a classic human face where traditional antennas are dovetailed, such as a stereotypical representation.

The two representations (Figure 6.5 and Figure 6.6) are classical and childish robots. They are squared with legs, arms, and a human-cartoon face. People who have drawn them could not specify their skills and

Figure 6.3 A Representation of a Robot. Author: Brazilian Artist, Thirthy Year, Man. Image Provided by the Author.

Figure 6.4 A Representation of a Robot. Author: Brazilian Film Maker, Twenty-Nine Years, Man. Image Provided by the Author.

competencies because they probably have just reproduced the classic stereotypical robot representation without thinking about its ontology and characteristics.

This woman-like robot (Figure 6.7) reminds Jetsons's robot, an anthropomorphic robot who serves a family. The robot has a name and a belly screen. The eyes and the screen are red indicating that they are electronic

The following drawing (Figure 6.8) is an odd representation of a robot. It has a human face with eyes and teeth. It has two antennas but also two "ears." When asked about these two ears, the little boy said that it is for "remembrancity"—actually, he refered to "remembrances." Asking if this "remembrancity" was linked to a brain, he argued that it was linked to "thought," but not necessarily to the brain, "because you may have thought without a brain." This statement reveals a close understanding of what AI

Figure 6.5 A Representation of a Robot. Author: Brazilian Artist, Thirty Years, Man. Image Provided by the Author.

may be. The classic squared body, articulated legs, and buttons are also reproduced in this representation.

The first drawing (Figure 6.9) has many dimensions. First, it regards a biped robot with a single articulated "tweezers hands." Curiously, it has a mobile instead of a shaped head. It is an iPhone with two persons on the screen in a virtual conversation. The drawer said that this robot who could react from the decisions made by the two people that are on the screen. The robot has an extra wide-angle lens for capturing the images.

The drawing (Figure 6.10) also has tweezer's hands and a pair of wheels. However, it has a analogical shape. Instead of the Iphone, it has an iPhone.

Figure 6.6 A Representation of a Robot. Author: Brazilian Artist, Thirty Years, Man. Image Provided by the Author.

A young Japanese boy made the drawing 6.11. At odds over the Western representation, the perception of a Japanese child brings emotional features to the robot. It has a sad face and a heart in the chest. Although we have observed human-like traits in the previous drawings, such as eyes, mouth, sometimes hair and hands, none of them expressed subjectivity as we see in the Japanese representation.

Another Japanese child lovingly represented the robot 6.12. By associating with her favorite character—a unicorn—she demonstrated that a robot is a friendly, sweet, and cute being/thing. It is worth saying that this drawing also has a heart, such as the former one. The anthropomorphic traits appear

Figure 6.7 A Representation of a Robot. Author: Brazilian Sociologist, Thirty-Two Years, Woman. Image Provided by the Author.

in the aesthetic details, such as the eyelashes and the neckless. It is awkward to observe that many of the drawings have old-fashioned representations of robots, mostly with antennas, which are not useful in cutting-edge technologies. Noteworthy to say that the robots here represented are either female or neutral. None has remarkable male features; instead, some have lips or long hairs.

From this quick and superficial exercise, we may draw some insights regarding how lay Western and Eastern people conceive AI. There is an evident emotive bias in Japanese drawings. They suggest that machines have emotions, not only intelligence. This Japanese empathetic connection to machines with intelligence must be a bit further explored.

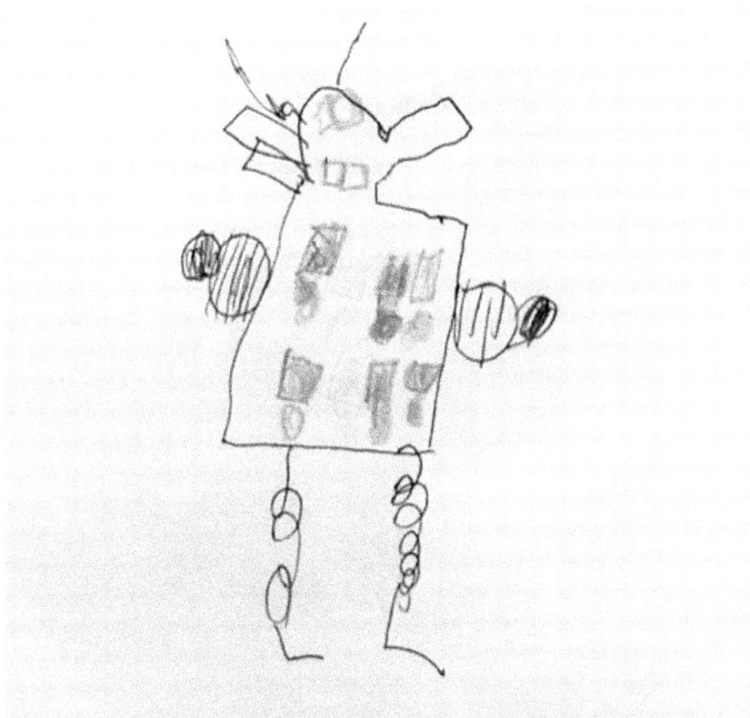

Figure 6.8 A Representation of a Robot. Author: Brazilian Student, Six Years, Boy. Image Provided by the Author.

TAKING THE FEARLESS TO THE NEXT LEVEL: THE JAPANESE FASCINATION FOR ROBOTS

If you are wandering around the streets of Tokyo, you will probably realize that you face a society in symbiosis with machines. There are game machines for entertainment, robots helping tourists with directions, food machines everywhere, among other details. The population aging made Japan to sell more diapers for elderly people than for children. By this token, machines try to compensate the shortage of labor force in the country. Instead of fearing machines, the Japanese became crazy for them, especially robots, which have been integrated into their social life in the past decades. Most of the machines have anthropomorphic traits that made them welcomed. More than that, they are a token of a more comfortable life. That is the reason why Japan is nowadays the leading producer of robots in the entire world. Nevertheless, what is the history and reasons for this love story? Why do we find a truly

Figure 6.9 A Representation of a Robot. Author: Brazilian Political Scientist. Thirty-Two Years, Man. Image Provided by the Author.

"robophilia" in Japan instead of a "robophobia" as we see in most parts of the West?

The positive and friendly attitude that the Japanese have toward robots is related not only to their socioeconomic context but also to their cultural context. First, the Japanese understand that the employment of robots helps their society to tick onward despite the decreasing birth rate. Robots are used in healthcare to substitute caregivers of children and elderly people (i.e., NEC's Papero robot). Also, there are robots for therapeutic communication (i.e., AIST's Paro), entertainiment (i.e., AIST's dinosaurs, Toyota's humanoids robots) or cleaning (i.e., Matsushita Electric, Fuji Heavy Industries). The social robot Pepper (that costs U$ 10.000) is already used in thousands of commercial places and houses across Japan. At odds over Western societies, the Japanese understand that robots are there for them, not to take things away. In this sense, Japan automation is less about threatening jobs and more about economic survival. In 2014, Prime Minister

Figure 6.10 A Representation of a Robot. Author: Brazilian Artist, Twenty-Nine Years, Man. Image Provided by the Author.

Shinzo Abe unveiled reforms hoping that the robot market reach $21 billion by 2020. Now, the unleashed incident of the COVID-19 pandemic and the social distancing measures, which may become habits, fostered the use of robots more than before. Thus, there is high competition among technology companies to develop a manifold of robots with different functions and traits.

Japan is not the only country that suffers from labor force shortage. However, at odds with Canada, for instance, which encourages immigration to deal with this issue, Japan is hesitant to accept immigrants. This attitude brings problems on a global level but mitigates many issues that they could have at a local level, such as slave labor, citizenship issue, and values and cultural clashes. All these

Figure 6.11 A Representation of a Robot. Author: Japanese Student, Six Years, boy. Image Provided by the Author.

issues become dissolved in the Japanese context because there are more in which robots occupying human functions in a very welcoming way.

This easy acceptance of robots in Japanese society is not only related to the economic but also the cultural domain. Japan developed the Otaku culture, the culture of *animés* and *mangás* where robots are represented as heroes.

After the World War II, from the ashes of the atomic bomb, Japan started to rebuild the country under the Western influence. In this context, in the 1950s, *Tetsuwan Atomu* (known as *Astro Boy*) raised from the creations of Osamu Tezuka. Astro Boy is a watershed in the history of robot culture because he became a national hero who lived in a future where robots and humans effortlessly embrace each other in a technology-human merge.

Figure 6.12 A Representation of a Robot. Author: Japanese Student, Four Years, Girl. Image Provided by the Author.

The cartoon romanticized the relation robot-human in that Astro Boy was built by a scientist to substitute his dead son.

Definitely, the process of robot's anthropomorphization is key to understand how Japanese society could solve the paradoxical relation human-technolog smoother than Western societies. Japan's robophilia might date back to the 1600s when *Karakuri ningyo* (automat or mechanized small puppets) were used for entertainment. While Western societies turned Pinnochios to fearing Robocops or Terminators, Japan turned *Karakuri* to cute and friendly Astro Boys.

These friendly traits helped the Japanese society to incorporate robots on a daily basis more naturally. In factories, you can already find pictures of robots among other human employee's pictures in a working map of a given area. Such situation might be weird, childish, creepy, or primitive for westerns. Otherwise, the Japanese established a symbiotic relationship not only with humanized machines but also with all the objects in the world. This attitude is based upon some premises of Shintoism, the most ancient religion in Japan.

Shintoism's animistic approach understands that the world is dwelt by inanimate things that have a soul, such as trees, stones, or objects. By this token, they might see things as having a sort of inner spark of life. Contrariwise, in the Judeo-Christian monotheistic tradition, only God can conceive life. Thereby, creating life is a direct threat to the conception of God itself. More than that, having a subjectivity means that this entity may have the same status as human beings. However, Shintoism is more about the relation between humans and things than conceiving subjectivity to objects. In this sense, the belief of spiritual life cannot be misunderstood as the idea of a robot with consciousness and subjectivity, such as explored in Western Science fiction. The "spirit" of an object in Japan is identified with its owner. When a robot interacts with its owner for a long time, it becomes attached and harmonized with its owner's spirit. Everything is fused without distinction between man, God, and the world. In this regard, aside from the public use of robots in Japan, they have been widely used in private environments, such as small or single-person households. They fill these void environments by having an essential social function in a country with a high aging population and immigrant rejections.

In this light, the Japanese industry, government, and academia have been creating cutting-edge robotic technologies along with a coherent and persuading narrative in which robots appear as pseudo-humans, who do things *for* humans, not *instead of* humans. Thus, they are seen firstly as "*controllable*" and secondly as "*artificial*" humans.

DO WE FEAR AI ITSELF, ITS CONSEQUENCES, OR ITS SHAPES?

AI does not only carry unforeseen risks; its main property may be a risk itself—depending on how people perceive AI substituting humans, either a solution or a hazard. On the one hand, in contexts where AI is needed because of the lack of humans, it may be deemed as a solution. On the other hand, in contexts where AI is not an immediate need, it may be deemed as a potential risk. With the contingent of immigrants in the United States, for instance, it is not straightforward to defend the employment of robots for services or production. However, Japan face off another scenario. There is a pride of being a homogeneous nation built around strict immigration policy. Immigrants compose only 1.2 percent of the population living in Japan. In 2018, the population fell by 430 thousand people. Thus, there are 1.63 jobs available for each citizen. The immigration issue has been declared and publicly debated for a long time in Japan. It has happened from ancient history to the present day. An example of that is the statement by the Japanese prime minister Yasuhiro Nakasone in 1986, saying that the intellectual level in the United States was lower than in Japan because there is a large population of African-American and Hispanics. Even if Japan had a more open immigration policy, its populated territory, the fact of being an island and its idiom would anyway hinder a sufficient migratory flow.

In this manner, the shortage of people in Japan creates a social problem that could be mostly overcome by robots, which could provide services, labor force in factories, social interaction, and entertainment. This makes the robot's acceptance more than a reality, a real need. They are everywhere, in many shapes and functions, performing a manifold of activities. In a nursing care home, baby-shaped robots wrapped in fabric diapers have been used to interact with elderly people. The same happens in a household to fulfill the void lagged by the only child who left home. There, household robots are deemed as family members who slide around the house.

AI bridges the gap between the need for the labor force and the amount of available workers.

Looking at the variety of shapes acquired by robots over Japan, it is worthy to note how the external features could cause much more societal rumors and impact than the technology inside itself. Sofia, the most famous robot in the world, is an example of that. Her media impact mostly owes to her realistic woman's look. The double robot of Hiroshi Ishiguro (the director of the Intelligent Robots Lab of the University of Osaka) is another astonishing example. As we said, Japanese society is more opener to robots, so, in this case, the robot truly looks like his own creator Ishiguro often sends his double to join talks and lectures in his place. Especially in Western countries, people usually get startled by the external format of Ishiguro's double. The fact that a robot could replace a human in a likely most human activity (in this case, to teach or deliver a talk) disturb people's image of what is "be a human"—at least in the Western. Thes creepy face of Ishiguro's robots portrays his face of ten years ago. When asked if he would build another robot who looks older to follow his aging, Ishiguro answers that he will rebuild his face to always look like his double.

This statement drew a lot of attention. Probably because there is a "sense of God" in what he says. Can we overcome the inevitable, the time passing? The point here is not AI anymore. The format and appearance of robots sometimes surpasses the debate about AI itself. Here, the issue is plastic art, not technology. Again, the representation of AI is always the first layer accessed by people, and its shape usually distracts the discussion of real technological development.

By analyzing the drawings of "artificial intelligence," we observe that people usually represent robots as anthropomorphic machines, but not like real people. A human's shape may cause strangeness or repellency for some cultures, but perhaps empathy and reliability for others. Much of the discussion around AI is still stuck on its shapes and external appearance. While there is little variation in how AI performs among its machines, there is tremendous variation in their outer forms, from its representation through non-anthropomorphized screens to perfect human images, such as Sofia or Ishiguro's double.

An insight into the forms often turns into real AI development and its consequences. Unlike many technologies, usually AI has its key objectives/functions and achievements merged. For example, a car has a primary objective to transport people faster, but as a result it can cause traffic or kill someone. That is, a car has an unintended consequences design.[1] In the case of AI, its main aim is to facilitate humans tasks or to intermediate (or bias) a human–human relationship—often through algorithms. Therefore, its consequences will be unemployment by substituting people's work. In this sense, AI inevitably involves politics and produces power and authority. With the spread of AI production in the world, it cannot be said that AI entails control over someone or some nation, as it happened with other technologies, such as the atomic bomb in the World War II. As AI has a manifold of applications and is spread everywhere, it is hard to detect what and where AI is precise. Slowly, AI silently reshapes our relations because "consciously or unconsciously, deliberatively or inadvertently, societies choose structures for technology that influence how people are going to work, communicate, consume."[2]

NOTES

1. Langdon Winner, *The Whale and the Reactor: A Search for Limits in an Age of High Technology* (Chicago: University of Chicago Press, 1989).
2. Winner, *The Whale and the Reactor*, 28.

Final Remarks

The representation of artificial intelligence as a forbidden fruit appeared in old speculations, for instance, in what the mysterious writer Ammianus Marcellinus portrayed in his short story *The Thought Machine*, for the science-fiction magazine *Amazing Stories* in the 1920s. Since then, this story has been retold over the years. There is a remarkable difference in the more recent versions of this story because the boundaries of fictional future scenarios and reality have been blurred. The development of technology contributed to that by filling the gap of what "we really have" in the present, and what we could have, would have, and will have. Conceiving possible future scenarios of artificial intelligence has become progressively dependent on the experts, who may be either scientists working on real applied research or intellectuals dedicated to thinking on conceptual and consequential issues related to that. Thereby, the teller of the stories has changed. We did not know who was the writer of the first stories on the rise of intelligent machines and their significant social consequences—something our illustrious unknown, Ammianus Marcellinus, did wittily. Instead, today, experts go public to profess future-imagined scenarios.

Even though most of the future scenarios conceived and conveyed are imaginary and hypothetical, it does not mean that the threats are impossible to happen or that we should not try to prevent undesired consequences of our technological undertaking. The point is that the process of conceiving these scenarios resembles exactly the process of building social-science fiction, which is the counterfactual thinking of possible worlds affected by the introduction of a new piece of technology. Throughout the periods we covered in this book, we have seen the narrowing of the gap between the real developed technology and the imagined consequences of the further advancements of it. Some experts and risk communicators present possible futures of imagined consequences and present facts of the actual technology as belonging to the same realm of possibility. Of course, the ease of availability and diffusion of contents on the internet enables the fast convergence of narratives, blurring any sharp distinction between what we have now and what we can have in the future (as well as its related consequences). In this context, experts and communicators are invited to exercise their social-scientific fiction skills professionally, but without the need to be fiction writers. The difference now is that these chains of imagined scenarios are self-proclaimed as nonfiction. The reason why I am paying attention to that is that within the context of fiction, we have the suspension of the misbelief. However, it does not downgrade the efforts of all these people. However, it shows the opposite, the laborious task of thinking on significant social economic and even psychological consequences of the development and use of any technology, in this case, artificial intelligence. This technology requires the flexibility of thinking a wide variety of scenarios since it is incredibly abstract, and its uses may be pervasive throughout different areas of economy and social life. Today, just a small part of that has a "human recognizable face," such as robots, but also intelligent assistants and chatbots, which can be anthropomorphized easily.

Certainly, the entertainment demand that we face today in our online access to contents is not only exclusive for cultural production but also and increasingly for intellectual and scientific production. An apparent side effect of that is the cast of doubts on the importance or the seriousness of some of these claims. The longer the links in the chains of these future scenarios are, the more farfetched they may look like. This side effect was criticized in this book at some points. However, these claims' underlying message is real. Any technological device we create will (potentially) bring us unexpected harms, and the extent of undesired consequences can be a behemoth. Prudence and risk management tells us that it makes sense, and it is worth to spend some resources thinking about how the scenarios with undesired consequences may look and how they may unfold. Still, all the future narratives are fragile because one cannot foresee all the links one needs to achieve a specific scenario and how weak (probabilistic speaking) the whole scenario might be.

It also creates an anomaly in how the perception of risk is built, amplified, and propagated through a certain amount of time within a given society. As we saw, the social amplification framework for the risk perception starts with an underlying, unequivocal harmful event, be it by death, injuries, or substantial financial losses. In the risk perception of artificial intelligence, the real underlying event misses, so it is replaced by the work of experts and intellectuals conceiving those scenarios. This is a hypothetical abstraction that paves the way to the infusion of (social) science-fiction thinking and pieces such as short stories, long stories, comics, and mainly movies. This is a major trend the risk perception of artificial intelligence in the past years was based on. We know now that the memorability of past events influences risk judgments and the imaginability of future events. As a result, any factor that makes a hazard unusually memorable or imaginable (in our case), such as a recent disaster, massive media coverage, or a vivid film, could seriously distort perceptions of risk. For us, movies, TV shows, and magazines that portray the behavior of full-fledged artificial intelligence and the undesired scenarios shape this perception of risk. Consequently, the risk of artificial intelligence is inherently epistemological because of its virtual (not real) existence. Dangers are not unreal, but most of the risks are still hypothetical. During the diffusion or propagation of the risk of artificial intelligence, reporters and other media players had no other options than trust in experts. We cannot forget that some experts who came to the forefront to communicate some artificial intelligence risks, such as Russell, are very prestigious scholars. In this sense, it is tough to disregard their positioning.

As we saw here, this story of the risk perception of artificial intelligence was shaped in the cyclical cultural tripod "representation—real scientific development—experts' public positioning." A crucial factor in the representation of the risk scenarios is the cultural influence. This is not a new argument, as we showed with the work of Mary Douglas, however that is not also the single factor that explains risk perception. Acknowledging the importance of the cultural influence of the representation of risk, we could perceive that there might exist a significant difference in the behavior disposition and the narratives between the Western countries, in particular the United States, and in an Eastern country, such as Japan.

Roughly, we presented some factors for the Japanese positive relationship with artificial intelligence, whether embodied or not. There is an older historic disposition, which can be traced back to Japan's pre–Modern era, related to the beliefs and behavior toward the suprasensible realm drawn by the Shintoism. In a nutshell, it enabled a non-suspicious relationship toward non-human entities that may have a few human traits. After the World War II, intelligent machines were positively represented, mainly because of the lack of people, destroyed families, and weakened workforce. The most striking example was the Atom Boy, a friendly robot created to substitute the son of a scientist. In the meantime, in the United States and the United Kingdom, robots were villains. It was interesting to explore that intelligent machines are integrated into Japanese society, not only narratively but also daily. There, artificial intelligence bridges the gap between the lack of people and the need of people.

The story of the risk perception is endlessly opened. The most recent chapter of the story, the 2020s., will draw on the popularization of AI and, likely on the attenuation of its potentially harmful effects. Many practical actions regarding specific applications of artificial intelligence have been taken, especially regarding the use of data and the bias in the dataset used in deep learning models. Thus, the actual real-world of artificial intelligence's side effects have been addressed with real-world solutions. However, there is still room for conceiving future scenarios, especially after the pandemic. Why? It is simple; it has its role. Such relative independence from the current underlying development, limitations, and usage of the technology shed light on the social, intellectual enterprise of conceiving future scenarios with undesired consequences. The first kick-off for the process of formation, propagation with amplification or attenuation of risk perception, and their societal, cultural, and economic impacts (which would be the real underlying event) may be less important than previously thought. Social-science fiction (which is more social-science fiction than social science-fiction) has played a pivotal role in other perceptions of risk we have. If it is the case, the cycle of representation, novelty (in this case, the real scientific and technological advancement), and communication (experts' public positioning) could also be verified. Still, in those future risk scenarios, the elements of thrill and adventure are always there.

Bibliography

Adorno, Theodor, and Max Horkheimer. *Dialectic of Enlightenment*. Stanford: Stanford University Press, 2002.

Alhakami, Ali, and Paul Slovic. “A Psychological Study of the Inverse Relationship between Perceived Risk and Perceived Benefit.” *Risk Analysis* 14, no. 6 (1994): 1085–96.

Anandan, Tanya. “Robot Safety, Everything But Routine.” *Robotic Industries Association* (blog), August 20, 2015. https://www.robotics.org/content-detail.cfm/Industrial-Robotics-Industry-Insights/Robot-Safety-Everything-But-Routine/content_id/5653.

Asimov, Isaac. “Social Science Fiction.” In *Modern Science Fiction: Its Meaning and Its Future*, edited by Reginald Bretnor. New York: Coward-McCann, 1953.

Atkins, B. T. S. “Building a Lexicon: The Contribution of Lexicography.” *International Journal of Lexicography* 4, no. 3 (1991): 167–204.

Baert, Patrick. “Positioning Theory and Intellectual Interventions.” *Journal for the Theory of Social Behaviour* 42, no. 3 (2012): 304–24.

Baert, Patrick, and Marcus Morgan. “A Performative Framework for the Study of Intellectuals.” *European Journal of Sociological Theory* 21, no. 3 (2017): 322–339.

Bailly, Christian. *Automata: The Golden Age: 1848–1914*. London: Robert Hale, 2003.

Barratt, James. *Our Final Invention: Artificial Intelligence and the End of the Human Era*. New York: Thomas Dunne Books, 2013.

Baum, Frank. *The Ozma of Oz*. Chicago: Reilly & Lee, 1907.

BBC. “The General Purpose Robot is a Mirage.” *Controversy*. London, 1973.

Beck, Ulrich. *Risk Society: Towards a New Modernity*. Translated by M Ritter. London: Sage, 1992.

Benson, Michael. *Space Odyssey: Stanley Kubrick, Arthur C. Clarke, and the Making of a Masterpiece*. New York: Simon & Schuster, 2019.

Bernstein, Peter. *Against the Gods: The Remarkable Story of Risk*. New York: John Wiley & Sons, 1996.

Bleiler, Everett. *Science Fiction: The Gernsback Years*. Kent, OH, and London: The Kent State University Press, 1998.

Bloomfield, Brian. *The Question of Artificial Intelligence: Philosophical and Sociological Perspectives*. London: Routledge Kegan & Paul, 1988.

Blount, Thomas. *Glossographia: Or A Dictionary, Interpreting All Such Hard Words of Whatsoever Language, Now Used in Our Refined English Tongue with Etymologies, Definitions, and Historical Observations on the Same. Also the Terms of Divinity, Law, Physick, Mathematicks, and Other Arts and Sciences Explicated. Very Useful for All Such as Desire to Understand What They Read.* London: Newcombe for George Sawbridge, 1661.

Boholm, Max, Niklas Möller, and Sven Ove Hanson. "The Concepts of Risk, Safety, and Security: Applications in Everyday Language." *Risk Analysis* 36, no. 2 (2016): 320–38.

Bostrom, Nick. *Superintelligence: Paths, Dangers, Strategies*. Oxford: Oxford University Press, 2014.

———. "What Happens When Our Computers Get Smarter than We Are?" presented at the *TED2015*, March 2015. https://www.ted.com/talks/nick_bostrom_what_happens_when_our_computers_get_smarter_than_we_are.

Bretton, Benjamin. "We Need to Talk About." *The Guardian*, December 30, 2013. http://www.theguardian.com/commentisfree/2013/dec/30/we-need-to-talk-about-ted.

Butler, Samuel. "Darwin Among the Machines." *Christ Church Press*. June 13, 1863. http://nzetc.victoria.ac.nz/tm/scholarly/tei-ButFir-t1-g1-t1-g1-t4.html.

———. *Erewhon; Or, Over the Range*. London: Trübner and Ballantyne, 1872.

Cambridge Dictionary. "Cliffhanger." In *Cambridge Online Dictionary*. Cambridge: Cambridge University Press, 2008. https://dictionary.cambridge.org/pt/dicionario/ingles/cliffhanger.

Casti, John. "Exit Homo Sapiens, Stage Left." *Nature*, February 25, 1999.

Chikayama, T., and R. Kiyohara. "Parallel Inference System Research in the Japanese FGCS Project." In *Parallel Symbolic Computing: Languages, Systems, and Applications*, edited by R. H. Halstead and T. Ito, 337–51. Lecture Notes in Computer Science. Berlin: Springer, 1993.

Clarke, Arthur C. *2001: A Space Odyssey*. New York: New American Library, 1968.

———. *Profiles of the Future: An Inquiry into the Limits of the Possible*. New York: Holt, 1984.

"Clarke Predicts Internet and PC." *ABC*, May 29, 1974.

Clynes, Manfred, and Nathan Kline. "Cyborgs and Space." In *Astronautics* 5, no. 9 (1960).

Compagnone, Antonio. "The Reconceptualization of Academic Discourse as a Professional Practice in the Digital Age: A Critical Genre Analysis of TED Talks." *HERMES—Journal of Language and Communication in Business* 27, no. 54 (December 22, 2015): 49. https://doi.org/10.7146/hjlcb.v27i54.22947.

Cotgrave, R., and C. Hollyband. *A Dictionarie of the French and English Tongues*. London: Adam Islip, 1611.

Crevier, Daniel. *AI: The Tumultuous History of the Search for Artificial Intelligence.* New York: Basic Books, 1993.

Douglas, Mary. *Purity and Danger: An Analysis of the Concepts of Pollution and Taboo.* London: Routledge, 1966.

Douglas, Mary, and Aaron Wildavsky. *Risk and Culture: An Essay on the Selection of Technological and Environmental Dangers.* Berkeley: University of California Press, 1982.

Dreyfus, Hubert. "Alchemy and Artificial Intelligence." Santa Monica, CA: The RAND Corporation, December 1965. http://www.rand.org/pubs/papers/2006/P3244.pdf.

———. *What Computers Can't Do: A Critique of Artificial Reason.* Cambridge, MA: MIT Press, 1972.

Du Cange, Charles. *Glossarium Ad Scriptores Mediae et Infimae Latinitatis, in Quo Latina Vocabula Novatae Significationis, Aut Usus Rarioris, Barbara & Exotica Explicantur, Eorum Notiones & Originationes Reteguntur.* Lutetiae Parisiorum: Ludovicum Billaine, 1678.

Dubarle, Dominique. "Une Nouvelle Science : La Cybernétique—Vers La Machine à Gouverner ?" *Le Monde*, December 28, 1948. http://sniadecki.wordpress.com/2013/03/08/dubarle-machine/.

Easton, Tom. "Robot: Mere Machinet Trancendent Mind Hans Moravec." *Analog: Science Fiction and Fact*, April 1999.

Edgerton, David. *The Shock of the Old: Technology and Global History since 1900.* London: Profile Books, 2006.

Eugene Rosa, Noriyuki Matsuda, and Randall Kleinhesselink. "The Cognitive Architecture of Risk: Pancultural Unity or Cultural Shaping?" In *Cross-Cultural Risk Perception: A Survey of Empirical Studies*, edited by Ortwin Renn and Bernd Rohrmann. Dordrecht: Springer Science+Business Media, 2000.

Farahany, Nita. "When Technology Can Read Minds, How We Will Protect Our Privacy?" Presented at the TED Salon: Zebra Technologies, November 2018. https://www.ted.com/talks/nita_farahany_when_technology_can_read_minds_how_will_we_protect_our_privacy.

Feigenbaum, Edward, and Pamela McCorduck. *The Fifth Generation.* Reading, MA: Addison-Wesley, 1983.

Fein, Louis. "Computer-Oriented Peace Research." *AFIPS '63* (Fall, 1963), 631–39.

———. "The Artificial Intelligentsia." *IEEE Spectrum* (February 1964), 74–87.

Fiksel, Joseph, and Vincent Covello. "Knowledge Systems, Expert Systems, and Risk Communication." In *Expert Judgement and Expert Systems*, edited by Mumpower et al., Vol. F35. NATO ASI Series. Berlin Heidelberg: Springer-Verlag, 1987.

Finucane, Melissa, Ali Alhakami, Paul Slovic, and Stephen Johnson. "The Affect Heuristic in Judgements of Risk and Benefits." In *The Perception of Risk*, edited by Paul Slovic, 413–29. Risk, Society and Policy Series. New York: Earthscan, 2000.

Fischhoff, Baruch, Paul Slovic, Sarah Lichtenstein, Stephen Read, and Barbara Combs. "How Safe Is Safe Enough? A Psychometric Study of Attitudes Toward Technological Risks and Benefits." In *The Perception of Risk*, edited by Paul Slovic, 413–29. Risk, Society and Policy Series. New York: Earthscan, 2000.

Florio, John. *A Vvorlde of Wordes, or Most Copious, and Exact Dictionarie in Italian and English, Collected by Iohn Florio.* London: Arnold Hatfield, 1598.

Forrester, Jared, Thomas Weiser, and Joseph Forrester. "An Update on Fatalities Due to Venomous and Nonvenomous Animals in the United States (2008–2015)." *Wilderness & Environmental Medicine* 29, no. 1 (March 2018): 36–44.

Foucault, Michel. *Discipline and Punish.* New York: Vintage Books, 1979.

———. "Governmentality." In *The Foucault Effect: Studies in Governmentality*, edited by G. Burchell, C. Gordon, and P. Miller, 87–104. London: Harvester Whearsheaf, 1991.

———. "Governmentality. Ideology and Consciousness." In *Power/Knowledge: Collected Interviews and Other Essays 1971–1977.* Brighton: Harvester Press, 1980.

Freudenburg, William. "Perceived Risk, Real Risk: Social Science and the Art of Probabilistic Risk Assessment." *Science* 242, no. 4875 (1988): 44–9. https://doi.org/10.1126/science.3175635.

Fuchi, Kazuhiro, Robert Kowalski, Koichi Furukawa, Kazunori Ueda, Ken Kahn, Takashi Chikayama, and Evan Tick. "Launching the New Era." *Communications of the ACM* 36, no. 3 (1993): 49–100.

Galeon, D. "Dubai Just Appointed a "State Minister for Artificial Intelligence." *Futurism*, October 20, 2017. https://futurism.com/dubai-just-appointed-a-state-minister-for-artificial-intelligence.

Garrick, B. John, Michael Greenberg, and Yacov Haimes, ed. "Stanley Kaplan: Reflections and Papers Published." *Risk Analysis* Special Issue (2019).

Gershenfeld, Neil. *When Things Start to Think.* New York: Henry Holt & Company, 1999.

Gilovich, Thomas, Dale Griffin, and Daniel Kahneman, ed. *Heuristics and Biases: The Psychology of Intuitive Judgment.* New York: Cambridge University Press, 2002.

Goffman, Erving. *The Presentation of Self in Everyday Life.* New York: Anchor Books, 1959.

Good, Irving J., "Speculations concerning the first ultraintelligent machine." Adv Comput 6:31–88, 1965.

Grimm, Jacob, and Wilhelm Grimm. *Deutsches Wörterbuch.* Leipzig: Hirzel, 1854.

Gruber, Tom. "Https://Www.Ted.Com/Talks/Tom_gruber_how_ai_can_enhance_our_memory_work_and_social_lives." presented at the TED2017, April 2017.

Hacking, Ian. *The Taming of Chance.* Cambridge: Cambridge University Press, 1990.

Halevy, Alon, Peter Norvig, and Fernando Pereira. "The Unreasonable Effectiveness of Data." *IEEE Intelligent Systems* (April 2009): 8–12.

Hall, Mordaunt. "A Technical Marvel." *The New York Times*, March 7, 1927. https://www.nytimes.com/1927/03/07/archives/a-technical-marvel.html?rref=collection%2Fcollection%2Fmovie-guide.

Hanson, Robin. *The Age of Em: Work, Love and Life When Robots Rule the Earth.* Oxford: Oxford University Press, 2016.

———. "What Would Happen If We Upload Our Brains to Computers?" Presented at the *TED2017*, April 2017. https://www.ted.com/talks/robin_hanson_what_would_happen_if_we_upload_our_brains_to_computers.

Hanson, Sven Ove. *The Ethics of Risk: Ethical Analysis in an Uncertain World.* Basingstoke: Palgrave Macmillan, 2013.

Hardy, Geofrey Harold. *A Mathematician's Apology.* Cambridge: Cambridge University Press, 2004.

Harris, Sam. "Can We Build Ai without Losing Control over It?" *TEDSummit*, June 2016. https://www.ted.com/talks/sam_harris_can_we_build_ai_without_losing_co ntrol_over_it.

———. "Science Can Answer Moral Questions." Presented at the TED2010, February 2010. https://www.ted.com/talks/sam_harris_science_can_answer_mora l_questions.

Hawking, Stephen. Stephen Hawking is Terrified of Artificial Intelligence. Interview by C. Kolodny. *Huffington Post*, May 5, 2014. https://www.huffingtonpost.co.uk/ entry /stephen-hawking-artificial-intelligence_n_5267481.

———. Stephen Hawking Warns Artificial Intelligence Could End Mankind. Interview by Rory Cellan-Jones. *BBC*, December 2, 2014. https://www.bbc.com/ news/technology-30290540.

Hillis, Daniel. Daniel Hillis. Interview by Christopher Sykes and George Dyson, October 2016. https://www.webofstories.com/play/danny.hillis/1.

Hohenemser, Christoph, Robert Kates, and Paul Slovic. "The Nature of Technological Hazard." In *The Perception of Risk*, edited by Paul Slovic, 168–81. Risk, Society and Policy Series. New York: Earthscan, 2000.

Homer. *The Odyssey.* Translated by Alexander Pope. London: Bernard Lintot, 1725–6.

Hornyak, Timothy. *Loving the Machine: The Art and Science of Japanese Robots.* Tokyo: Kodansha International, 2006.

House of Commons. "Robotics and Artificial Intelligence." *House of Commons*, 2016. https://publications.parliament.uk/pa/cm201617/cmselect/cmsct ech/145/14502. htm.

House of Lords. "AI in the UK: Ready, Willing and Able?" *House of Lords*, 2018. https://www.parliament.uk/business/committees/committees-a-z/ lords-select/ai -committee/news-parliament-2017/ai-report-published/.

Iyer, Ravishankar. "'Faultproof and Incapable of Error?' Reliable Computing and Fault Tolerance." In *HAL's Legacy: 2001's Computer as Dream and Reality*, edited by David Stork. Cambridge, MA: The MIT Press, 1997.

Joy, Bill. "Why the Future Does Not Need Us?" *Wired*, December 1999, 124–35.

Kahneman, Daniel. *Thinking, Fast and Slow.* New York: Farrar, Straus and Giroux, 2011.

———. "Varieties of Counterfactual Thinking." In *What Might Have Been: The Social Psychology of Counterfactual Thinking*, edited by Neal Roese and James Olson. London: Psychology Press, 1995.

Kahneman, Daniel, Paul Slovic, and Amós Tversky, ed. *Judgment under Uncertainty: Heuristics and Biases.* New York: Cambridge University Press, 1982.

Kahneman, Daniel, and Amós Tversky. "The Simulation Heuristic." In *Judgment under Uncertainty: Heuristics and Biases*, edited by Daniel Kahneman, Paul Slovic, and Amós Tversky. Cambridge: Cambridge University Press, 1998.

Kant, Immanuel. *Critique of Practical Reason*. Translated by Mary Gregor. Cambridge: Cambridge University Press, 2015.

Kaplan, Stanley. "On The Method of Discrete Probability Distributions in Risk and Reliability Calculations–Application to Seismic Risk Assessment." *Risk Analysis* 1, no. 3 (1981): 189–96.

———. "The Words of Risk Analysis." *Risk Analysis* 17, no. 4 (1997): 407–17.

Kaplan, Stanley, and B. John Garrick. "On The Quantitative Definition of Risk." *Risk Analysis* 1, no. 1 (1981): 11–27.

Kapur, Arnav. "How AI Could Become an Extension of Your Mind." Presented at the *TED2019*, April 2019. https://www.ted.com/talks/arnav_kapur_how_ai_could_become_an_extension_of_your_mind.

Kasperson, Roger, Ortwin Renn, Paul Slovic, Halina Brown, Jacque Emel, Robert Goble, Jeanne Kasperson, and Samuel Ratick. "The Social Amplification of Risk: A Conceptual Framework." In *The Perception of Risk*, edited by Paul Slovic, 232–45. Risk, Society and Policy Series. New York: Earthscan, 2000.

Kedrowicz, April A., and Julie L. Taylor. "Shifting Rhetorical Norms and Electronic Eloquence: TED Talks as Formal Presentations." *Journal of Business and Technical Communication* 30, no. 3 (July 2016): 352–77. https://doi.org/10.1177/1050651916636373.

Knight, Frank. *Risk, Uncertainty, and Profit*. Boston and New York: Houghton Mifflin Co., 1921.

Kracauer, Siegfried. *From Caligari to Hitler: A Psychological History of German Film*. Princeton, NJ: Princeton University Press, 1947.

Kuck, David. "Could We Build HAL? Supercomputer Design." In *HAL's Legacy: 2001's Computer as Dream and Reality*, edited by David Stork. Cambridge, MA: The MIT Press, 1997.

Kurzweil, Ray. *The Age of Intelligent Machines*. Cambridge, MA: MIT Press, 1990.

———. *The Age of Spiritual Machines: When Computers Exceed Human Intelligence*. New York: Viking, 1999.

Lanier, Jaron. "DIGITAL MAOISM: The Hazards of the New Online Collectivism." *Edge.Org*, May 29, 2006. https://www.edge.org/conversation/digital-maoism-the-hazards-of-the-new-online-collectivism.

———. "How We Need to Remake the Internet." Presented at the *TED2018*, April 2018. https://www.ted.com/talks/jaron_lanier_how_we_need_to_remake_the_internet.

———. "The Myth of AI: A Conversation with Jaron Lanier." *Edge.Org*, November 14, 2014. https://www.edge.org/conversation/the-myth-of-ai#26015.

———. *You Are Not a Gadget: A Manifesto*. New York: Vintage Books, 2010.

Lee, Kai-Fu. *AI Superpowers: China, Silicon Valley, and the New World Order*. Boston, MA: Houghton Mifflin Harcourt, 2018.

Lewis, David. "Truth in Fiction." *American Philosophical Quarterly* 15, no. 1 (1978): 37–46.

Licklider, J. "Man-Computer Symbiosis." *IRE Transactions on Human Factors in Electronics,* HFE-1 (1960): 4–11.

Lighthill, James. *Artificial Intelligence: A Paper Symposium*. London: Science Research Council, 1973.

Ludewig, Julia. "TED Talks as an Emergent Genre." *CLCWeb: Comparative Literature and Culture* 19, no. 1 (March 1, 2017). https://doi.org/10.7771/1481-4374.2946.

Luhmann, Niklas. *Risk: A Sociological Theory*. New York: A. de Gruyter, 1993.

Lyng, Stephan. "Crime, Edgework, and Corporeal Transaction." *Theoretical Criminology* 8, no. 3 (2004): 359–75.

———. "Disfunctional Risk Taking: Criminal Behavior as Edgework." In *Adolescent Risk Taking*, edited by N. Bell and R. Bell. London: Sage, 1993.

———. "Edgework: A Social Psychological Analysis of Voluntary Risk Taking." *American Journal of Sociology* 95 (1990): 851–86.

———, ed. *Edgework: The Sociology of Risk-Taking*. New York: Routledge, 2005.

Machlis, Gary, and Eugene Rosa. "Desired Risk: Broadening the Social Amplification of Risk Framework." *Risk Analysis* 10 (1990): 161–68.

Marcellinus, Ammianus. "The Thought Machine." *Amazing Stories*, February 1927.

Mateas, Michael. "Reading HAL: Representation and Artificial Intelligence." In *Stanley Kubrick's 2001: A Space Odyssey. New Essays*, edited by Robert Kolker. Oxford: Oxford University Press, 2006.

Mayor, Adrienne. *Gods and Robots: Myths, Machines and Ancient Dreams of Technology*. Princeton, NJ: Princeton University Press, 2018.

McAfee, Andrew. "Are Droids Taking Our Jobs?" Presented at the *TEDxBoston2012*, Boston, June 2012. https://www.ted.com/talks/andrew_mcafee_are_droids_taking_our_jobs.

———. "What Will Future Jobs Look Like?" Presented at the *TED2013*, February 2013. https://www.ted.com/talks/andrew_mcafee_what_will_future_jobs_look_like/transcript.

McAleer, Neil. *Sir Arthur C. Clarke: Odyssey of a Visionary*. New York: RosettaBooks LLC, 2013.

McCarthy, John. John McCarthy (1927–2011): Artificial Intelligence. Interview by J Mishlove, 1989. https://thinkingallowed.com/2jmccarthy.html.

———. John McCarthy Oral History. Interview by Nils Nilsson, September 12, 2007. https://www.computerhistory.org/collections/catalog/102658149.

———. "Review of 'Artificial Intelligence: A General Survey' by Professor Sir James Lighthill." *Artificial Intelligence* 5, no. 3, 1974.

———. "Review of The Question of Artificial Intelligence Edited by Brian Bloomfield." *Annals of the History of Computing*, 1989.

———. "What is Artificial Intelligence." November 12, 2007. http://www-formal.stanford.edu/jmc/whatisai.pdf.

McCorduck, Pamela. *Machines Who Think: A Personal Inquiry About the History and Prospects of Artificial Intelligence*. Natick: A K Peters, 2003.

McGinn, Colin. "Hello, HAL." *New York Times*, January 3, 1999. https://archive.nytimes.com/www.nytimes.com/books/99/01/03/reviews/990103.03mcginnt.html.

Merton, Robert. "The Unanticipated Consequences of Purposive Social Action." *American Sociological Review* 1, no. 6 (1936): 894–904.

Michael, Chris. Spike Jonze on Letting Her Rip and Being John Malkovich. *The Guardian*, September 9, 2013. https://www.theguardian.com/film/filmblog/2013/sep/09/spike-jonze-her-scarlett-johansson.

Mike Ashley. *The Time Machines: The Story of the Science-Fiction Pulp Magazines from the Beginning to 1950*. Vol. 1. Liverpool: Liverpool University Press, 2000.

Milgram, Anne. "Why Smart Statistics Are the Key to Fighting Crime." Presented at the TED@BCG, San Francisco, October 2013. https://www.ted.com/talks/anne_milgram_why_smart_statistics_are_the_key_to_fighting_crime.

Minsky, Marvin. Kurzweil Interviews Minsky: Is Singularity Near? Interview by Ray Kurzweil, 2010. https://www.kurzweilai.net/kurzweil-interviews-minsky-is-singularity-near.

———. Marvin Minsky. Interview by Christopher Sykes, January 31, 2011. https://www.webofstories.com/play/marvin.minsky/52.

———. *The Emotion Machine: Commonsense Thinking, Artificial Intelligence, and the Future of the Human Mind*. New York: Simon & Schuster, 2006.

Moravec, Hans. "Harvard Doesn't Publish Science Fiction New Destinies, Volume III, Jim Baen, Ed., Baen Books, February 1988, Pp. 212–247." In *New Destinies*, edited by Jim Baen, III:212–47. Baen Books, 1988.

———. "Intelligent Machines: How to Get There from Here and What to Do Afterwards." September 1977. https://frc.ri.cmu.edu/~hpm/project.archive/general.articles/1977/smart.

———. "Rise of the Robots." *Scientific American*, December 1999, 124–35.

———. *Robot: Mere Machine to Transcendent Mind*. New York: Oxford University Press, 1998.

———. "The Age of Robots." In *Proceedings of the First Extropy Institute Conference on TransHumanist Thought*, edited by Max More, 84–100. Sunnyvale: Extropy Institute, 1994.

———. "The Endless Frontier and The Thinking Machine, in The Endless Frontier, v2, Jerry Pournelle, Ed., Grosset & Dunlap, Ace Books, January 1982, Pp. 374–397. (Invited)." In *The Endless Frontier*, edited by Jerry Pournelle, 2:374–97. New York: Ace Books, 1982.

Neri, Hugo. "The Process of Intellectualization: Foundations for a Sociological Explanation of Knowledge." University of São Paulo, 2014. https://teses.usp.br/teses/disponiveis/8/8132/tde-13082014-210252/publico/2014_HugoNeriMunhoz_VOrig.pdf.

Neri, Hugo, and Fabio Cozman. "The Role of Experts in the Public Perception of Risk of Artificial Intelligence." *AI & Society* 35, no. 3 (2020), 663–673. https://doi.org/10.1007/s00146-019-00924-9.

NHTSA, National Highway Traffic Safety Administration. "Traffic Safety Facts: Research Note." U.S. Department of Transportation, September 2012. https://crashstats.nhtsa.dot.gov/Api/Public/ViewPublication/811650.

Nietzsche, Friedrich. *Thus Spoke Zarathustra*. Translated by Walter Kaufmann. New York: Random House, 1954.

Nilsson, Nils. *The Quest for Artificial Intelligence: A History of Ideas and Achievements*. Cambridge: Cambridge University Press, 2009.

NSTC. "National Science and Technology Council Committee on Technology. Preparing for the Future of Artificial Intelligence." NSTC—Executive Office of the President National Science and Technology Council, 2016. https://obamawhitehouse.archives.gov/sites/default/files/whitehouse_files/microsites/ostp/NSTC/preparing_for_the_future_of_ai.pdf.

Oakley, Brian, and Kenneth Owen. *Alvey: Britain's Strategic Computing Initiative*. Cambridge, MA: MIT Press, 1989.

Oxford Dictionary. "Cliffhanger." In *Oxford Dictionary of English*. Oxford: Oxford University Press, 2019.

———. "Futurist." In *Oxford Dictionary of English*. Oxford: Oxford University Press, 2019.

———. "Futurology." In *Oxford Dictionary of English*. Oxford: Oxford University Press, 2019.

———. "Risk." In *Oxford Dictionary of English*. Oxford: Oxford University Press, 2019.

Pearl, Judea. *Causality: Models, Reasoning, and Inference*. 2nd ed. Cambridge: Cambridge University Press, 2009.

———. *The Book of Why: The New Science of Cause and Effect*. New York: Basic Books, 2018.

Pidgeon, Nick, Roger Kasperson, and Paul Slovic, ed. *The Social Amplification of Risk*. Cambridge: Cambridge University Press, 2003.

Pierini, Francesco. "The Popularization of Specialized Knowledge Through Ted Talks: The Case of Positive Psychology." *International Journal of English Linguistics* 9, no. 4 (June 12, 2019): 15. https://doi.org/10.5539/ijel.v9n4p15.

Platt, Charles. "Redundancy Has Its Virtues." *Wired Magazine*, December 1998. https://frc.ri.cmu.edu/~hpm/book97/reviews/981201.wired.html.

Poole, David, and Alan Mackworth. *Artificial Intelligence: Foundations of Computational Agents*. 2nd ed. Cambridge: Cambridge University Press, 2017.

Pooley, Jefferson, and Michael Socolow. "The Myth of the War of the Worlds Panic." *Slate*, October 28, 2013. https://slate.com/culture/2013/10/orson-welles-war-of-the-worlds-panic-myth-the-infamous-radio-broadcast-did-not-cause-a-nationwide-hysteria.html.

Quandt, T., L. Frischlich, and T. Schatto-Eckrodt. "Fake News." In *The International Encyclopedia of Journalism Studies*, edited by T. P. Vos, F. Hanusch, D. Dimitrakopoulou, M Geertsema-Sligh, and A. Sehl. New Jersey: John Wiley & Sons, 2019.

Rees, Martin. *Our Final Hour: A Scientists Warning; How Terror, Error, and Environmental Disaster Threaten Humankind's Future in This Century – on Earth and Beyond*. New York: Basic Books, 2003.

Renn, Ortwin. "Concepts of Risk: A Classification." In *Social Theories of Risk*, edited by S. Krimsky and D. Golding, 53–79. Westport, CT and London: Praeger, 1992.

———. "Risk Perception and Risk Management: A Review." *Risk Abstracts* 7, no. 1 (1990).

Renn, Ortwin, William Burns, Jeanne Kasperson, Roger Kasperson, and Paul Slovic. "The Social Amplification of Risk: Theoretical Foundations and Empirical Applications." *Journal of Social Issues* 48, no. 4 (1992): 137–60.

Rhodes, Richard. *The Making of the Atomic Bomb*. New York: Simon & Schuster, 1986.

Rieder, John. *Science Fiction and the Mass Cultural Genre System*. Middletown: Wesleyan University Press, 2017.

Roberts, Adam. *The History of Science Fiction*. 2nd ed. London: Palgrave Macmillan, 2016.

Roble, Doug. "Digital Humans That Look Like Us." Presented at the *TED2019*, April 2019. https://www.ted.com/talks/doug_roble_digital_humans_that_look_just_like_us.

Rohrmann, Bernd, and Ortwin Renn. "Risk Perception Research—An Introduction." In *Cross-Cultural Risk Perception: A Survey of Empirical Studies*, edited by Ortwin Renn and Bernd Rohrmann. Dordrecht: Springer Science+Business Media, 2000.

Romanelli, Frank, Jeff Cain, and Patrick J. McNamara. "Should TED Talks Be Teaching Us Something?" *American Journal of Pharmaceutical Education* 78, no. 6 (August 2014): 113. https://doi.org/10.5688/ajpe786113.

Rosenblatt, Frank. *Principles of Neurodynamics*. Washington DC: Spartan Books, 1962.

———. "The Perceptron: A Probabilistic Model for Information Storage and Organization in the Brain." *Psychological Review* 65 (1958): 386ff.

Russell, Bertrand. *The Scientific Outlook*. London: Routledge, 2001.

Russell, Stuart. "3 Principles for Creating Safer AI." Presented at the *TED2017*, April 2017. https://www.ted.com/talks/stuart_russell_3_principles_for_creating_safer_ai.

———. "Ban Lethal Autonomous Weapons." *The Boston Globe*, August 9, 2015.

———. *Human Compatible: Artificial Intelligence and the Problem of Control*. New York: Viking, 2019.

———. "Take a Stand on AI Weapons." *Nature* 521 (2015): 415–16.

Russell, Stuart, and Peter Norvig. *Artificial Intelligence: A Modern Approach*. 3rd ed. Upper Saddle River: Prentice Hall, 2009.

Sankar, Shyam. "The Rise of Human-Computer Cooperation." *TEDGlobal 2012*, June 2012. https://www.ted.com/talks/shyam_sankar_the_rise_of_human_computer_cooperation.

Schank, Roger, and Robert Abelson. *Scripts, Plans, Goals and Understanding: An Inquiry into Human Knowledge Structures*. New Jersey: Erlbaum, 1977.

"Sci-Fi at CHI: Cyberpunk Novelists Predict Future User Interfaces." In *CHI' 92* (1992): 435–37.

Scotto di Carlo, Giuseppina. "Ethos in TED Talks: The Role of Credibility in Popularised Texts." *Facta Universitatis*, Linguistics and Literature, 12, no. 2 (2014): 81–91.

———. "New Trends in Knowledge Dissemination: TED Talks." *Acta Scientiarum. Language and Culture* 36, no. 2 (May 9, 2014): 121. https://doi.org/10.4025/actascilangcult.v36i2.22619.

Selfridge, Oliver, and Ulrich Neisser. "Pattern Recognition by Machine." *Scientific American* 203 (1960): 60–8.

Simmel, Georg. "The Sociology of Secrecy and of the Secret Societies." *American Journal of Sociology* 11 (1906): 441–98.

Simon, Herbet. *Models of Man*. London: John Wiley & Sons, 1957.

Slovic, Paul. "Informing and Educating the Public about Risk." *Risk Analysis* 6, no. 4 (1986): 403–15.

———. "Introduction and Overview." In *The Perception of Risk*, edited by Paul Slovic. Risk, Society and Policy Series. New York: Earthscan, 2000.

Slovic, Paul, Melissa Finucane, Ellen Peters, and Donald MacGregor. "The Affect Heuristic." *European Journal of Operational Research* 1777 (2007): 1333–52.

Slovic, Paul, Baruch Fischhoff, and Sarah Lichtenstein. "Cognitive Processes and Societal Risk Taking." In *The Perception of Risk*, edited by Paul Slovic, 32–50. New York: Earthscan, 2000.

———. "Rating the Risks." *Environment* 2, no. 3 (1979): 14–20.

Slovic, Paul, Howard Kunreuther, and Gilbert White. "Decision Processes, Rationality and Adjustment to Natural Hazards." In *The Perception of Risk*, edited by Paul Slovic, 1–31. Risk, Society and Policy Series. New York: Earthscan, 2000.

Smith, John Maynard. John Maynard Smith. Interview by Richard Dawkins, April 1997. https://www.webofstories.com/play/john.maynard.smith/1.

Starr, Chauncey. "Social Benefit versus Technological Risk." *Science* 165 (1969): 1232–38.

Stefanik, Elise. "H.R. 5356 to Establish the National Security Commission on Artificial Intelligence." 2018. https ://www.congress.gov/bill/115th-congress/house-bill/5356.

Stilgoe, Jack. "Machine Learning, Social Learning and the Governance of Self-Driving Cars." *Social Studies of Science* 48, no. 1 (2017): 25–56.

Stork, David. "Scientist on the Set: An Interview with Marvin Minsky." In *HAL's Legacy: 2001's Computer as Dream and Reality*, edited by David Stork. Cambridge, MA: The MIT Press, 1997.

Suarez, Daniel. "The Kill Decision Shouldn't Belong to a Robot." *TEDGlobal2013*, June 2013. https://www.ted.com/talks/daniel_suarez_the_kill_decision_shouldn_t_belong_to_a_robot.

Sugg, Stewart. *Slaughterbots*. Space Digital, 2017. https://www.youtube.com/watch?v=9CO6M2HsoIA.

Suvin, Darko. *Positions and Presuppositions in Science Fiction*. London: MacMillan Press, 1988.

Tegmark, Max. *Life 3.0: Being Human in the Age of Artificial Intelligence*. New York: Knopf, 2017.

———. "Max Tegmark: How to Get Empowered, Not Overpowered, by AI." Presented at the *TED2018*, April 2018. https://www.ted.com/talks/max_tegmark_how_to_get_empowered_not_overpowered_by_ai.

The English Reports. *The English Reports*. London: W. Green, 1900.

Turing, Alan. "Computing Machinery and Intelligence." *Mind* LIX, no. 236 (1950): 433–60.

Tversky, Amós, and Daniel Kahneman. "Judgment under Uncertainty: Heuristics and Biases." *Science* 185 (1974): 1124–31.

Waldrop, Mitchell. "The Souls of the New Machines." *New York Times*, January 1, 1989. https://frc.ri.cmu.edu/~hpm/book88/reviews/890101.NYT.html.

Weber, Max. *Max Weber-Gesamtausgabe (MWG): Wirtschaft Und Gesellschaft. Die Wirtschaft Und Die Gesellschaftlichen Ordnung Und Mächte. Nachlaß. Religiöse Gemeinschaften,* edited by Hans Kippenberg and Petra Schilm, Vol. 22–2. Max Weber-Gesamt Ausgabe, I. Tübingen: Mohr Siebeck, 2001.

Westfahl, Gary. "Introduction: Of Futures Imagined, and Futures Inhabited." In *Science Fiction and the Prediction of the Future: Essays on Foresight and Fallacy*, edited by Gary Westfahl, Wong King Yuen, and Amy Kit-sze Chan. Critical Exploration in Science Fiction and Fantasy 27. Jefferson, NC: Mc Farland & Co., 2011.

Wiener, Norbert. *Cybernetics; or, Control and Communication in the Animal and the Machine*. Paris: Technology Press, 1948.

———. *God & Golem, Inc.: A Comment on Certain Points Where Cybernetics Impinges on Religion*. Cambridge, MA: MIT Press, 1964.

———. "Some Moral and Technical Consequences of Automation." *Science* 131 (1960): 1355–58.

———. *The Human Uses of Human Beings*. 2nd ed. London: Free Association Books, 1989.

Winner, Langdon. *The Whale and the Reactor: A Search for Limits in an Age of High Technology*. Chicago: University of Chicago Press, 1989.

Workman, Christopher, and Troy Howarth. *Tome of Terror: Horror Films of the Silent Era*. Baltimore: Midnight Marquee Press, 2016.

Yudkowsky, Eliezer. "Artificial Intelligence as a Positive and Negative Factor in Global Risk." In *Global Catastrophic Risk*, edited by Nick Bostrom and Milovic Cirkov. Oxford: Oxford University Press, 2008.

Zamith-Cruz, Judith, and Paulo Vieira. "Can Machine Intelligence Overcome the Human One?" In ed. Manuel Curado and Steven Gouveia. *Automata's Inner Movie: Science and Philosophy of Mind*. Wilminton: Vernon Press, 2019.

Zinn, Jens, ed. *Social Theories of Risk and Uncertainty: An Introduction*. Oxford: Blackwell Publishing, 2008.

Zuboff, Shoshana. *The Age of Surveillance Capitalism: The Fight for a Human Future at the New Frontier of Power*. New York: PublicAffairs, 2019.

Index

About the Author

Hugo Neri is a researcher at the Center for Artificial Intelligence (IBM/University of São Paulo/São Paulo Research Foundation) and postdoc at the Department of Mechatronics Engineering and Mechanical Systems of the University of São Paulo. He is a lecturer at the Sociology Department, Innsbruck Universität. He was a visiting scholar at the Department of Sociology of the University of Cambridge, the Goethe Universität and the Knowledge Lab/University of Chicago. He holds a PhD degree in philosophy by the University of São Paulo, Master's in sociology and B.A. in social sciences by the University of São Paulo. His research focuses on the foundation, representation, and application of commonsense and risk perception among philosophy, sociology, and computer sciences.

www.ingramcontent.com/pod-product-compliance
Ingram Content Group UK Ltd.
Pitfield, Milton Keynes, MK11 3LW, UK
UKHW020842290926
15470UKWH00004B/1492

9 781793 602053